Handbook of Hazardous

Waste

Management

for Small
Quantity Generators

RUSSELL W. PHIFER
WILLIAM R. McTIGUE, JR.

 LEWIS PUBLISHERS

Library of Congress Cataloging-in-Publication Data

Phifer, Russell W.
  Handbook of hazardous waste management for small quantity
generators.

  Bibliography: p.
  Includes index.
  1. Hazardous wastes—United States—Management—
Handbooks, manuals, etc. 2. Hazardous wastes—Law and
legislation—United States—Handbooks, manuals, etc. I. McTigue,
William R. II. Title.
TD811.5.P47   1988   628.5   87-29678
ISBN 0-87371-102-5

Fourth Printing 1990

Third Printing 1989

Second Printing 1988

LEWIS PUBLISHERS, INC.
121 South Main Street, Chelsea, Michigan 48118

PRINTED IN THE UNITED STATES OF AMERICA

To Colleen

Preface

While it is the belief of the authors that all information supplied in this book is accurate, regulatory definitions and interpretations are constantly changing. The information supplied represents our best effort to interpret regulations and does not necessarily represent the opinions or legal interpretations of any regulatory agency, organization, or anyone other than the authors. If the reader notices any errors, the authors would appreciate being advised so that future editions may be corrected.

The authors wish to thank the following organizations and individuals for their valuable assistance in this project.

Special Hazards Management Company, Inc.
Envirite Field Services, Inc.
American Chemical Society
U.S. Environmental Protection Agency
U.S. Department of Transportation
RCRA Superfund Hotline
EPA Small Business Ombudsman
Individuals at the various state environmental agencies
Fitch & Sandell, Inc.
Richard D. Trimpi
Jeffrey S. Fitch
Errol S. Fletcher

Russell W. Phifer

Mr. Phifer has specialized in the management of small quantity generator wastes since 1980. As Chairman of the American Chemical Society's Task Force on RCRA, he is involved in establishing policies on
hazardous waste management for laboratories and other small generators of waste. Other responsibilities include the development of
training programs and publications on hazardous waste management.
Mr. Phifer has served as a guest lecturer at a number of colleges and
universities, and has advised a number of state and federal regulatory
agencies on waste management policies and regulations. He has published technical papers on laboratory waste management, laboratory
safety, risk assessment, and the problems of small quantity generators. Mr. Phifer is a graduate of the College of Wooster (Ohio) and
currently serves as an Environmental Planner for Fitch & Sandell,
Inc. in Wayne, Pennsylvania. He is a Certified Hazardous Materials
Manager. Mr. Phifer is married and resides in Mont Clare,
Pennsylvania.

William R. McTigue, Jr.

Mr. McTigue has served in the hazardous waste management industry in numerous capacities, including: manager of a commercial hazardous waste storage facility; project coordinator in emergency response actions; permit administrator for commercial TSD facilities; environmental regulatory consultant; health and safety manager for a hazardous waste remediation firm. His experience has required him to interact extensively with governmental bureaus, dealing particularly with enforcement agents and permit writers at the federal and state levels. His approach to regulatory problem solving is based firmly in understanding a situation in the context of applicable regulations and designing a cost-effective solution which is sensitive to a business firm's resource base. Mr. McTigue received his BA from Temple University and his MBA from La Salle University in Philadelphia. Approved by the Institute of Hazardous Materials Management as a Certified Hazardous Materials Manager (CHMM), he is currently the manager of health and safety for Envirite Field Services, Inc., Plymouth Meeting, Pennsylvania.

Contents

ix

List of Figures

List of Tables

1

Why You Need This Book

It is difficult to imagine any issue in the '80s more visible than the protection of our environment. We all share a common concern over the quality of life, and one of the most obvious mistakes of our society is the way we have traditionally approached the disposal of waste. The technical advances which are largely responsible for our consumer society should have also provided ways to properly manage the by-products of this industrial explosion. It should be one of our greatest regrets that industry did not seriously consider the issue of hazardous waste management until it was forced to do so by a suddenly awakened public.

It was primarily public pressure that forced Congress to begin passing environmental legislation at a rapid pace, beginning with the Resource Conservation and Recovery Act (RCRA) in 1976. RCRA gave the U.S. Environmental Protection Agency (EPA) the responsibility of regulating the disposal of hazardous waste, with a relatively short time to write and promulgate regulations that would have a profound effect on industry.

In the beginning, the EPA had to consider the most effective ways to initiate responsible hazardous waste management by industry, without presenting an impossible financial burden. Since 10% of the nation's businesses generated over 95% of the waste, it was decided to begin by concentrating on these large businesses. This was accomplished by exempting those businesses generating less than 1000 kg (2200 lb) in any one given month from most of the regulations.

Where does this leave small businesses? Until 1986, small quantity generators, assuming they were able to find disposal sites to handle their waste, could avoid most of the requirements for recordkeeping, training, reporting, manifesting, and a number of other regulations. Is there a substantial burden placed on small business to comply with the additional requirements introduced under the RCRA reauthorization? Since in most cases it is up to the individual states to implement their own laws, the answer to this question varies. The sheer number of different agencies with which businesses must contend to properly and legally manage their wastes can be overwhelming even to large companies with small armies of environmental compliance workers. Indeed, there is little evidence of cooperation among the different regulatory agencies, and compliance with the regulations in one state may result in violations in another.

Who are the small generators most affected by additional regulation? Many are not obvious—they range from high schools to bowling alleys to pharmacies to dry cleaners. Some trade associations and other organizations are assisting in providing the necessary information and direction. In many more cases, small generators are just now waking up to the serious implications of new federal and state regulations involving the management of hazardous wastes. It is for these generators that this book is intended. With costs for disposal rising between 10% and 50% per year for many waste streams, any serious effort to reduce costs and manage wastes more efficiently can make a big difference for both large and small generators.

The dollar cost of proper disposal has surprised many small generators, largely because they do not believe they are generators and, in fact, are not obviously in that category. Secondary academic institutions are a perfect example. Nearly every school in the country has a chemistry program, and the past practice of maintaining a large chemical storeroom has resulted in quantities of unused, out-of-date, and otherwise surplus chemicals. In the past many high schools have been delighted to accept free chemicals from a laboratory going out of business or just trying to reduce inventory. High schools have sometimes worked this procedure in reverse, but current disposal costs and the slow realization that these chemicals may end up being waste in the future has resulted in a decrease in the off-site recycling of laboratory chemicals.

Is there one great solution to the problem of hazardous waste management for small generators? An emphatic *no* is the answer, but there are a number of small answers that can add up to a successful program. Minimization of off-site disposal quantities, improved efficiency in operations, and reduced inventory and storage costs can result from a properly planned and managed program. These gains, though, come only with a program that takes advantage of all of the small solutions available.

What are the available off-site disposal options for hazardous waste? Geographical considerations are important when considering the available options. In the western states, for example, landfilling is still considered the main option, largely due to the lack of commercially available incinerator sites. The eastern states, however, have more options available because chemical and thermal treatment facilities compete with incinerators for disposal dollars.

Certainly one of the prime considerations is the cost of transportation to the various disposal sites. As more and more treatment, storage and disposal facilities (TSDFs) are closing, distance from a disposal site becomes significant. This should encourage the building of more regional sites, but public outcry against sites in their neighborhoods has resulted in the permitting of few sites since 1980, when RCRA was implemented. Few investors are willing to take the risk involved in trying to get permitted, given the large engineering and legal costs required well before a permit can be considered. It is standard for environmental activists to mobilize against any proposed facility, regardless of the level of planning, location, and need for new facilities. Small businesses need to recognize when new sites in their regions are necessary to help reduce costs and to participate in the siting process along with the environmental concerns. Only with the active participation of waste generators will there be any new sites permitted.

In this book we sift through the myriad requirements for waste management and provide small generators with the information to design a waste management program in compliance with current regulations. We encourage a high level of involvement by upper level management, since only with a strong commitment to the development and implementation of a well-designed waste management plan can small businesses successfully contend with both the legal and fundamental requirements of a successful program. The difference between good planning and poor planning is a survival issue for many small businesses; failure to meet the requirements of federal and state hazardous waste regulations can be costly in terms of fines and other sanctions.

2

Introduction to the Regulations

THE REGULATORY SCHEME

Many statutes, both federal and state, have been written to provide for the protection of human health and the environment. While they are intended to provide for such protection by addressing either specific or general issues, they must be amplified and implemented through various agencies of government. Despite their authors' highest aspirations and the best efforts of administering agencies, the statutes' objectives are not always achieved. Consequently, more statutes and regulations are written. Frequently, overlapping of these laws occurs. At times there is consistency and at other times there is conflict. Thus, compliance with one set of rules does not ensure compliance with another set. However, proceeding conscientiously within the letter and spirit of the law, one is apt to minimize his personal exposure to criminal and civil suits as well as to achieve the primary objective: protection of human health and the environment.

Key to achieving the goal of protecting human health and the environment is a working knowledge of applicable regulations. A general understanding of the law from which agencies derive their authority to promulgate and enforce regulatory standards is important, since changes in federal regulations impact significantly on a host of programs operated at the state level.

Federal Legislation

The regulation of hazardous waste is based on two major pieces of federal legislation: the Resource Conservation and Recovery Act of 1976 (RCRA) and the Hazardous and Solid Waste Amendments of 1984 (HSWA). To the extent that HSWA amends RCRA, it is safe to say that the federal hazardous waste management program is authorized by one federal law: RCRA. It is worth noting that RCRA's purview is not limited simply to hazardous waste management; it is actually directed at the regulation of all solid waste, a subset of which is hazardous waste. While the statute has numerous objectives — including technical and financial assistance to state and local governments for the development of solid waste management plans, resource recovery and conservation systems, and the prohibition of open dumping on the land — its basic intent is to provide for the protection of human health and the environment. To ensure that the regulated community shares its intent, Congress has provided for penalties of "not more than $50,000 per day of violation, or by imprisonment for not more than two years, or by both." [RCRA, Sec. 3008] These penalties are for the most serious violations. Obviously, the federal government views proper waste management quite seriously. Citizens are advised to do so as well.

The RCRA statute addresses various topics pertaining to solid waste in an organized manner, namely, delineating major subjects as subtitles. Each subtitle is divided into sections. The subtitle of RCRA which focuses on hazardous waste management is Subtitle C. However, other sections of the statute are also relevant to hazardous waste regulation. These are ultimately incorporated in the hazardous waste regulations established by the U.S. Environmental Protection Agency (EPA).

EPA Regulations

While RCRA provides general objectives and timetables, it does not detail the manner in which objectives are to be met. Instead, the law charges EPA with the responsibility of developing and implementing programs that are consistent with the intent of Congress. In effect, regulations promulgated by the EPA become federal law. All such standards developed by the agency are published in the *Federal Register* (FR). The FR is a publication of the United States government which reprints notices, proposed rules, and final rules issued by every major agency of the federal government. (The FR is published Monday through Friday throughout the year and is available on a subscription basis.)

Annually, the rules published in this periodical are summarized in a publication entitled *Code of Federal Regulations* (CFR). Each federal agency has its own title of CFR. For example, the EPA's title is "Protection of Environment," and it has been assigned the corresponding numeral 40. Thus, the citation of any EPA regulation will typically be cited as "40 CFR" followed by the relevant section number. Rules which are issued by the EPA subsequent to the latest revision of 40 CFR can be found in the FR. It should also be noted that references to subtitles and sections of RCRA will often be cited in the literature. These are codified in the CFR according to section numbers assigned by the EPA. While a literal reading of the statute will always be informative, it will generally not yield the specific information that evolves from the EPA's rulemaking process.

State Authorization of RCRA Programs

In addition to authorizing the EPA to administer the RCRA program, Congress also provided via the statute the opportunity for state governments to develop their own waste management systems. However, a state is not permitted to administer the RCRA program without first receiving permission from the EPA. States which administer their own hazardous waste management programs must demonstrate consistency with federal standards and must also meet specific criteria established by the EPA. Such state programs must be at least as stringent as the federal program. In fact, to the extent that its program is consistent with federally mandated objectives and standards, a state is permitted to institute a system which is more stringent than the federal program.

A state which is permitted to administer RCRA for a temporary period is said to have *interim* authorization (Phase I authorization). It is during this period that the state program is supervised and evaluated by the EPA. A state which operates under interim status has the authority to regulate waste management practices in its jurisdiction according to its own laws. One which ultimately meets EPA requirements is considered to have *final* authorization (Phase II authorization). This type of state is able to administer its program in lieu of the federal program. Such a state is also described as having *primacy*.

A state which administers its own RCRA program must develop its own laws and vest in a designated agency the authority to administer the laws' provisions. The revision, publication, and compilation of laws and regulations at this governmental level typically parallels the manner in which activities are conducted at the federal level. Each state develops its own regulations by statute and adapts them into some type of administrative code. Changes in regulations are published in a state periodical that is made available to the public through various means. Thus, changes which affect the state's regulated community can be referenced in various state codes and timely publications.

If the regulatory scheme seems relatively straightforward to this point, don't be fooled! There are still many aspects of the regulations which require further development by the EPA and interpretation by the courts.

The constant change of the regulations at the federal level necessarily disrupts the consistency between federal and state programs. (Remember, even states which have primacy must have a RCRA program that is consistent and at least as stringent as EPA's system.) A perfect example of this type of disruption is the change in the regulation of small quantity generators (SQGs) at the federal level. Until the change in the federal requirements, many states which had primacy regulated SQGs according to EPA's standards. However, as mandated by HSWA, in March 1985 the EPA promulgated regulations which substantially tightened control over SQGs. Until the states with primacy develop regulations which are consistent with the newer federal standards, SQGs must abide by federal regulations. On the other hand, there were other states which regulated SQGs much more stringently than the EPA prior to March 1985. Waste generators in these particular states, whether or not their state had interim or final authorization status, were subject to their particular state's controls.

While this inconsistency concept can be discussed exhaustively, the critical point is this: the first step in complying with RCRA is to identify the relevant issues and governmental jurisdictions. Ultimately, it is the generator's responsibility to comply with applicable federal and state regulations. This book will provide the necessary guidance to help you achieve compliance with the RCRA program.

RESPONSIBILITIES OF THE SMALL QUANTITY GENERATOR

In basic terms, compliance with environmental regulatory standards is accomplished by executing three steps:

- determining the jurisdiction in which the waste falls
- determining the quantity of waste generated
- evaluating the waste in terms of the regulations

Determination of Jurisdiction

Before attempting to identify whether a waste is a hazardous waste,

it is crucial to determine whether or not the waste generator is subject to the regulations of the federal program or those of an authorized state program.

The easiest method of making this determination is to take advantage of the information that the regulatory agencies provide. By following any one of the three steps listed below, anyone can readily determine whether he is subject primarily to the federal or state hazardous waste program.

- Call the RCRA Superfund Hotline.
- Contact the Hazardous Waste Management Branch of the regional office of the EPA for your geographical area.
- Call the state agency responsible for regulating solid waste disposal.

A list of names, addresses, and telephone numbers for all applicable state and federal agencies is provided in Appendix I of this handbook.

Remember, even if you operate in a state which has final authorization to administer RCRA, you must keep abreast of changes in the federal program. The promulgation of new regulations by EPA may cause inconsistencies between federal and authorized state programs.

Determining Quantity of Waste Generated

Generally speaking, as the quantity of hazardous waste (produced in a calendar month) increases, so do the demands of governmental regulation. In determining whether or not a generator produces a quantity of hazardous waste which causes him or her to be subject to regulation, a generator must understand EPA's rules concerning hazardous waste generation. It is useful to categorize generators in three ways:

- large quantity
- small quantity
- conditionally exempt

EPA's definition of a *small quantity generator* can be summarized as follows:

A small quantity generator is one who produces in a calendar month

- more than 100 kilograms (kg) but less than 1000 kg of nonacutely hazardous waste
- less than 100 kg of waste resulting from the cleanup of any residue or contaminated soil, water, or other debris involving the cleanup of an acutely hazardous waste
- less than 1 kg of an acutely hazardous waste*

*1 kilogram = 2.2 lb.

Generators who exceed these quantities are considered *large generators*. Generators who produce less than these quantities are considered *conditionally exempt generators*. That is, if they comply with certain conditions, they need not manage their wastes in accordance with the more stringent regulations established for small and large quantity generators of hazardous waste. An important point in determining the quantity of waste produced, however, is to recognize that certain exemptions and exclusions for types of waste must be considered. These and other issues are described in the section "Hazardous Waste Determination."

Hazardous Waste Determination

Understanding Definitions and Exclusions

After you have determined which program you are subject to (federal or state), follow the steps relating to solid waste and hazardous waste determination. These are stated and briefly discussed below. Recall that a waste can be hazardous only if it is first a solid waste.

1. *Comprehend the statutory definition of the term **solid waste** in order to develop a basic understanding of the scope of wastes that are subject to RCRA control.*

While many people might argue that this step is superfluous and provides little, if any, value in complying with specific requirements, the generality of the statutory definition is worth noting. Upon reviewing it, you will likely experience an overwhelming sense that your waste materials are subject to RCRA control. If the statutory definition is not enough to convince you, then be advised that the EPA has the authority (per sections of the RCRA and Comprehensive Environmental Response, Compensation and Liability Act [the Superfund law]) to determine that your waste is at the very least a solid waste and, at the most, a hazardous waste. The RCRA definition of a solid waste is

> . . . any garbage, refuse, sludge from a waste treatment plant, water supply treatment plant, or air pollution control facility and other discarded material, including solid, liquid, semisolid, or contained gaseous material resulting from industrial, commercial, mining, and agricultural operations, and from community activities, but does not include solid or dissolved material in domestic sewage, or solid or dissolved materials in irrigation return flows or industrial discharges which are point sources subject to permits under section 402 or the Federal Water Pollution Control Act, as amended (86 Stat. 880), or source, special nuclear, or byproduct material as defined by the Atomic Energy Act of 1954, as amended (68 Stat. 923.1). [RCRA, Sec. 1004, para. 27]

Hopefully, by now you are convinced of the extreme probability that you are generating a waste that is either subject to RCRA regulation or is specifically excluded per RCRA authority.

2. *Realize that the statute specifically* **excludes** *from the definition of solid waste several wastes that are generated from industrial processes.*

These exclusions are reprinted in Appendix II of this book. They are referenced in 40 CFR 261.4 (a)(1-7). The most important point to note is that

> if your activity or waste generating process falls within any of these exclusions, then your waste is not a solid waste and, therefore, not a hazardous waste.

3. *Besides those wastes excluded per RCRA section 1004 [as described in 40 CFR 261.4 (a)], note that* **additional exclusions** *are provided in that section.*

These exclusions are reprinted in Appendix III of this book. Of particular interest in the remainder of 261.4 is the exclusion pertaining to waste **samples**. The exclusion for samples facilitates the waste analysis process. However, one should pay keen attention to the specific conditions of the samples exclusion. Violations of these conditions would necessarily imply noncompliance with the regulations.

This section also excludes **household waste** from hazardous waste control. Household waste is not limited to those wastes generated by conventional single and multiple family residences, but can also include hotels, motels, ranger stations, and other residence situations. These residences frequently generate waste materials identical to those generated by regulated businesses, but they have been excluded by the EPA for a variety of reasons.

4. *Recognize that materials that are* **recycled** *in particular ways are not solid wastes.*

These are stated in 261.2 (e)(i-iii) as materials that are

a. used or reused as ingredients in an industrial process to make a product, provided the materials are not being reclaimed; or

b. used or reused as effective substitutes for commercial products; or

c. returned to the original process from which they are generated, without first being reclaimed. The material must be returned as a substitute for raw material feedstock, and the process must use raw materials as principal feedstocks.

Evaluation of Nonexcluded Wastes

If your waste is not specifically excluded as a solid or hazardous waste, then you must evaluate it according to the standards specified in 262.11. These standards are listed below.

Step 1. *Determine if the waste is listed in* **Subpart D** *of 40 CFR Part 261.*

Step 2. *Determine whether the waste is identified in* **Subpart C** *of 40 CFR Part 261 by either*:

a. "testing the waste according to the methods set forth in Subpart C of 40 CFR Part 261, or according to an equivalent method approved by the Administrator under 40 CFR 260.21"; or

b. "applying knowledge of the hazard characteristic of the waste in light of the materials or the processes used."

Each of these subparts is now outlined in detail.

I. **EPA Lists of Hazardous Wastes.** Subpart D of 40 CFR 261 consists essentially of four tables which list wastes, the processes generating them, and their corresponding EPA hazard classes and waste identification numbers. Basically, these tables identify (1) **hazardous wastes from nonspecific sources;** (2) **hazardous wastes from specific sources;** and (3) **discarded commercial chemical products, off-specification species, container residues, and spill residues, which are either acutely toxic or toxic.** These tables are listed in Appendix IV of this book. When working with the tables, one must pay particular attention to the **specific language** that describes their applicability to the waste determination process.

A. **Hazardous Wastes from Nonspecific Sources.** Commonly referred to as the **F List,** this table identifies a host of waste chemicals which are generated from any number of processes. It consists of three headings: (1) Industry and EPA Hazardous Waste Number, (2) Hazardous Waste, and (3) Hazard Code.

1. **Headings**

a. **Industry and EPA Hazardous Waste Number.** This column indicates the single appropriate **EPA identification number** for one or any number of wastes that are described in the adjacent column labeled "Hazardous Waste." This number must be used to describe the waste on the hazardous waste manifest and the EPA form "Notification of Hazardous Waste Activity."

b. **Hazardous Waste.** The information contained under this heading describes a general process which results in the generation of specific and generic chemical wastes. The key to using this table properly is to identify with particularity both the **process** which generates the waste **AND** the **constituents** of the waste. Then simply refer to the table and determine if your information matches any of the descriptions provided in this column. If it matches, your hazardous waste is described by the corresponding EPA number and hazard code. If it does not, you must consult the other tables and/or other steps of the waste determination process.

c. **Hazard Code.** This column indicates a **letter** that describes a generic class of hazard by EPA code. The hazard code, like the identification number described above, must be used on the hazardous waste manifest and the notification form. This hazard code is especially important in determining generator status, since it may describe a waste as being **acutely hazardous**. The applicability of conditional exemptions and SQG exclusions pivot on the quantity of acutely hazardous waste produced by a waste generator. Thus, the designation of (H) for a waste stream should be given special attention. **Multiple hazard classes** are occasionally assigned to a particular type of waste. In such cases, the waste must always be described by the multiple classes, since the generator has no discretion in hazard class assignment. (This statement does not apply to mixtures of solid and hazardous wastes.)

2. **Peculiarities.** The F Table pulls into RCRA control many types of waste that might otherwise be free from regulation. A couple of peculiarities of the regulation of wastes described by numbers F001 through F005 are worth noting.

a. With few exceptions, the **land disposal** of any hazardous waste which contains these wastes is prohibited. Fortunately, however, one of these exceptions applies to generators of 100–1000 kg of waste per month. This exception is valid until November 8, 1988. (Note: certain conditions apply to F003 wastes, as discussed below. Also, the land disposal of liquid hazardous wastes is strictly prohibited.)

b. The description of the F001-F005 wastes incorporates what is commonly known as the **solvent mixture rule**. (Further information on this rule can be found in the *Federal Register*, December 31, 1985.) The solvent mixture rule, which relates **only** to F001-F005 wastes, applies to waste determination in a very interesting way; it considers the concentration (by volume) of the presence of these chemical compounds in a solution **before** use. That is, the concentration of the chemicals in the ultimate waste product is meaningless. The applicability of the hazardous waste regulations to the spent solvent/solvent mixture hinges on the concentration of these chemicals in the material's virgin form. Thus, if prior to use, the waste material **was composed of 10% or more of a single chemical OR consisted of 10% or more of a mixture of these chemicals,** then it is subject to hazardous waste regulation by listing. Namely, the waste must be described by its corresponding waste number and the corresponding hazard code.

When examining the hazard codes that are associated with each of the F listings, notice that solvents listed as (1) **F001, F002, F004, and F005** are listed due to their toxicity (T); (2) **F003** are listed exclusively due to their ignitability (I); (3) **F005** must always be described by both of the hazard codes assigned: (I,T).

Therefore, any hazardous waste which is composed of 10% or more of any of the F solvents listed for toxicity (T) is always hazardous due to listing. The "10% concentration" refers to a total of 10% of any or all of the F solvents listed for toxicity. Therefore, **any hazardous waste that consists of 10% or more of the F solvents which are generally listed for toxicity (T), OR any solid waste that is mixed with these wastes will always be considered a hazardous waste and must always be described by the relevant F number and corresponding hazard code(s).**

Example. Consider wastes, each of which in their **unused** form consisted solely of water and the concentration (by volume) of the chemicals specified below. Table I classifies the waste according to applicable regulations.

Example. Consider a situation in which a shop proprietor who is subject to hazardous waste regulation stores several gallons of solvent in a drum outside his store. Suppose that he has determined that it meets the definition of an F005 waste. The drum springs a leak, and its contents contaminate the ground on which the drum is sitting. When the proprietor discovers the problem, he shovels the contaminated soil into a

Table I. Classification of Waste According to F List

Constituents	Waste Number	Hazard Code
10% carbon tetrachloride	F001	T
10% carbon tetrachloride, 10% toluene	F001, F005	I,T
5% carbon tetrachloride, 5% toluene	F001, F005	I,T
5% carbon tetrachloride, 4% toluene	NOT REGULATED BY LISTING	
9% carbon tetrachloride	NOT REGULATED BY LISTING	
3% carbon tetrachloride, 2% chlorobenzene, 3% cresylic acid, 2% methyl ethyl ketone	F001, F002, F004, F005	I,T

new container. The contaminated dirt is a hazardous waste. Its EPA identification number is F005 and its hazard code is (I,T). Clearly, the mixture is subject to hazardous waste regulation.

It is interesting to note that if the proprietor in the example were storing a waste solvent that did not bear a hazard code of (T) or (H), the contaminated soil would not be a listed hazardous waste as described in 40 CFR Part 261 Subpart C. The proprietor would then determine whether or not the waste exhibits any characteristics of hazardous waste. This concept will be discussed later. According to 40 CFR 261.3 (a)(2)(iii), if the mixture does not exhibit any such characteristics, then it is not a hazardous waste. This leads us to conclude that **a hazardous waste that is listed for toxicity or acute toxicity, whether it is mixed with solid waste or other hazardous waste, is subject to RCRA Subtitle C regulation. It must be described by its corresponding F number and hazard code and is thus regulated as a listed waste.**

3. **The F003 listing.** By now it is clear that any waste which meets the listing of F001, F002, F004, and F005 is a hazardous waste, and it bears the toxicity (T) hazard code. Whether the waste is mixed with a solid waste or not, it must always be described by the appropriate F number and toxicity (T) code.

In hazardous waste determination the applicability

of the F003 listing is less clear-cut. According to the EPA, a few criteria are considered in determining whether or not a waste meets this listing. These criteria are applied in the context of the solvent mixture rule; they consider the concentrations of the F compounds in solvents/solvent mixtures **before** use.

a. **Considerations**

  (1) In cases where a solvent mixture (prior to use) included 10% or greater of an F003 solvent and any of the other F solvents, then the waste is identified as F003 (I) and F00(1, 2, 4, or 5) (T).

  (2) In cases where a solvent/solvent mixture contained (prior to use) greater than 10% of any of the F003 solvents and less than 10% of any of the other F solvents, the solvent would **not necessarily** be an F003 waste. The key here is whether or not the F003 solvents were of a commercial grade. If they were of a commercial grade, then the F003 listing would certainly apply. If not, then the waste would only have to be evaluated for characteristics of hazardous waste. [40 CFR Part 261 Subpart C] These characteristics, described in 40 CFR Part 261 Subpart C are: ignitability (I), corrosivity (C), reactivity (R), and Extraction Procedure (EP) Toxicity (E).

  (3) In cases where a hazardous waste meets the listing of F003 and is mixed with a solid waste, AND where the solid/hazardous waste mixture does not exhibit any hazardous waste characteristics (I, C, R, or E), then the mixture is not a hazardous waste. [40 CFR 261.3 (a)(2)(iii)]

b. **Application.** A careful review of the language of the F003 listing provides insight into the apparent confusion of its applicability in the hazardous waste determination process. In excerpts from the listing, we see it addressing three general scenarios:

  (1) "the following spent nonhalogenated solvents: xylene, acetone, ethyl acetate, ethyl benzene, ethyl ether, methyl isobutyl ketone, n-butyl alcohol, cyclohexanone, and methanol"; [This

portion of the listing quite simply addresses not mixtures of the spent solvents, but only the spent form of their commercial grade.]

(2) "all spent solvent mixtures/blends containing, before use, only the above spent nonhalogenated solvents"; [This portion of the listing seems to address only mixtures and blends composed **exclusively** of these solvents.]

(3) "and all spent solvent mixtures/blends containing, before use, one or more of the above nonhalogenated solvents, and a total of 10% or more of those solvents listed in F001, F002, F004, and F005; and the still bottoms from the recovery of these spent solvents and solvent mixtures." [This portion of the listing effectively captures a broad range of solvent mixtures.]

Hence, the solvents covered by the F003 listing differ from other solvents insofar as they are listed, not for their toxicity, but for their ignitability.

B. **Hazardous Waste from Specific Sources.** As the description of the tables implies, all wastes that are listed here result from specifically described processes. All wastes identified in this table are common in one respect: their EPA waste number is prefixed by the letter **K**. Thus, these are typically called **K wastes.**

Like the F table (with the notable exception of F001 through F005), the K table regulates various wastes without regard to concentration. In general, whether a solid waste contains one part per million (ppm) or several hundred thousand ppm of a K waste, it will be a hazardous waste and is subject to Subtitle C regulation. The exceptions to this rule pertain to **K044, K045, and K047 wastes,** since they are listed solely for their characteristic of reactivity (R)—a characteristic that is described in 40 CFR Part 261 Subpart C. Recall that section 261.3 (a)(2)(iii) indicates that a mixture of a solid waste and a hazardous waste that is listed in Subpart D solely because it exhibits one or more of the characteristics of hazardous waste in Subpart C is not a hazardous waste if it (the mixture) does not exhibit any of those characteristics.

The K list is structured like the F list. It is composed of three columns: Industry and EPA Hazardous Waste Number, Hazardous Waste, and Hazard Code. Any waste that meets the listing must be described by the corresponding waste number and hazard code.

C. **Discarded Commercial Chemical Products.** While the F and K lists are the controls employed by EPA to pull into the hazardous waste management system a variety of spent chemicals from nonspecific and specific processes, they provide no basis for Subtitle C control of various unused discarded chemicals, off-specification chemical products, or manufacturing chemical intermediates. The lists which provide such a basis are generally referred to as the P and U tables.

1. **Types of information in the P and U tables.** The P and U tables are composed of two headings: Hazardous Waste Number and Substance. Their applicability in the hazardous waste determination process is described in 40 CFR 261.33. It is useful to consider 261.33 as containing two distinct types of information: (1) a set of **conditions**, and (2) general **descriptions** of materials/substances, which if they meet any of those conditions, are hazardous wastes. Naturally, the conditions and general descriptions must be used in conjunction with one another.

   a. **Conditions.** Any material described in 261.33 which is handled in any of the following ways is a hazardous waste:

      (1) discarded or intended to be discarded;

      (2) mixed with waste oil, used oil, or other materials, and applied to the land for dust suppression or road treatment; or

      (3) instead of being used as intended, produced for use as a fuel, or burned as a fuel.

   b. **Materials.** As stated in the regulations, the above conditions apply to these materials:

      (1) "any commercial chemical product, or manufacturing chemical intermediate having the generic name listed in [the P or U table],"

      (2) "any off-specification commercial chemical product or manufacturing chemical intermediate, which if it met specifications, would have the generic name listed in [the P or U table],"

(3) "any container or inner liner removed from a container that has been used to hold any commercial chemical product or manufacturing chemical intermediate having the generic names listed [in the P table only] . . . or any container or inner liner removed from a container that has been used to hold any off-specification chemical product and manufacturing chemical intermediate, which if it met specifications, would have the generic name listed in [the P table], unless the container is empty as defined in 261.7 (b)(3) of this chapter,"

(4) "any residue or contaminated soil, water, or other debris resulting from the cleanup of a spill into or on any land or water of any

(a) commercial product listed in [the P or U tables],

(b) manufacturing chemical intermediate listed in [the P or U tables], or

(c) off-specification chemical product or manufacturing chemical intermediate which, if it met specification, would have the generic name listed in [the P or U tables.]"

2. **Use of the P and U tables**
   a. **EPA comments.** Two comments in the regulations are provided by EPA concerning the use of these tables. The first, which deals with the residue of a P chemical in an empty container, notes that the residue is **necessarily** a hazardous waste **unless** it is being beneficially used, reused, legitimately recycled or reclaimed, or is being accumulated, stored, transported, or treated prior to such beneficial activities. In managing a container which previously held a P chemical, it is extremely important to understand the definition of an empty container. (This definition will be discussed shortly.)

   The second comment attempts to clarify the

meaning of the phrase "commercial chemical product or manufacturing chemical intermediate having the generic name listed in." EPA interprets the phrase as referring "to a chemical substance which is manufactured or formulated for commercial or manufacturing use which consists of (1) the commercially pure grade of the chemical, (2) any technical grades of the chemical that are produced or marketed, and (3) all formulations in which the chemical is the sole active ingredient." [Format changed for emphasis.]

EPA states clearly that these lists are **not** relevant in describing a manufacturing process waste that contains a substance listed in the P or U tables. Rather, such wastes must be evaluated in terms of the F and K lists or by the characteristics of hazardous waste that are described in Part 261 Subpart C.

b. **Analysis.** A careful, literal reading of this second comment yields a rather startling deficiency of EPA's regulation of P and U wastes. Namely, discarded formulations which contain **multiple** active ingredients, some or all of which are listed on the P or U tables, are **not** subject to hazardous waste regulation per the P or U listings. Given that such formulations are neither spent solutions nor manufacturing process wastes, they are not likely to be found on the F or K tables. Further, since the overwhelming majority of them are listed exclusively due to their toxicity (T) or acute toxicity (H) – characteristics which are not defined in Part 261 Subpart C – they fall out of the hazardous waste management system! Under no circumstances, however, is it advisable to intentionally blend wastes that would be regulated as P or U wastes to avoid compliance with regulatory standards.

Another point that is worth emphasizing, too, is the idea that ultimately the regulation of discarded commercial chemical products, intermediates, residues, and so on per the P and U tables typically is not based on the concentration of the respective P or U substance. For example, consider a discarded pesticide whose sole active ingredient is heptachlor. Whether the heptachlor is present in the formula in a concentration of 0.01% or 100%, the discarded pesticide is subject to regulation per P listing. Heptachlor's EPA waste number is P059 and its hazard code is (H). Four notable exceptions to this statement, however, include chemicals identified as P001, P122, U248, and U249. Nonetheless, whether the substance is above or below the concentrations stated by these waste numbers, it is still subject to hazardous waste regulation, provided that it meets the conditions and descriptions stated in 261.33.

The small quantity generator is urged to use great care in evaluating the types and quantities of wastes generated, particularly when dealing with substances described in the P table. It is advisable to include (in the monthly quantity of waste generated) the actual weight of all empty containers that do not meet the definition of **empty** as described in 40 CFR 261.7 (b)(3).

As discussed, while the F, K, P, and U tables describe a variety of wastes, they represent only a portion of the hazardous wastes that are generated in this country. Recognizing the mind-boggling variety of chemicals that are used in commerce, EPA realizes that it cannot possibly assemble a list that will subject to regulation all of those wastes which threaten human health and the environment. Therefore, the agency developed a concept which relates to hazardous characteristics which a waste might exhibit. Hence, if the generator has determined that his waste is not hazardous by listing, he must proceed to the second step of the hazardous waste determination process: determining whether the waste is identified in Subpart C of 40 CFR Part 261.

II. **EPA Characteristics of Hazardous Waste.** Subpart C of 40 CFR Part 261 identifies four characteristics of hazardous waste: (1) **Ignitability**, (2) **Corrosivity**, (3) **Reactivity**, and (4) **EP Toxicity**. Attempting to evaluate a solid waste for hazardous waste characteristics based on a poor knowledge of the materials or processes that were used to generate the waste invites trouble on all levels — legal, financial, health, and environmental. If any doubt exists about the hazardous characteristics of the wastes, enlist the services of a laboratory qualified to perform the tests prescribed in Subpart C by EPA. The regulatory definitions for each characteristic are reprinted below.

A. **Ignitability**

  1. **Definition.** A solid waste exhibits the characteristic of ignitability if a representative sample of the waste has any of the following properties:

     a. It is a liquid, other than an aqueous solution containing less than 24 percent alcohol by volume and has a flash point less than 60 degrees C (140 degrees F), as determined by a Pensky-Martens Closed Cup Tester, using the test method specified in ASTM Standard D-93–79 or D-93–80 (incorporated by reference, see Sec. 260.11), or a Setaflash Closed Cup Tester, using the test method specified in ASTM Standard D-3278–78 (incorporated by reference, see Sec. 260.11), or as determined by an equivalent test method approved by the Administrator under procedures set forth in Sec. 260.20 and 260.21.

     b. It is not a liquid and is capable, under standard temperature and pressure, of causing fire through friction, absorption of moisture or spontaneous chemical changes and, when ignited, burns so vigorously and persistently that it creates a hazard.

     c. It is an ignitable compressed gas as defined in 49 CFR 173.300 and as determined by the test methods described in that regulation or equivalent test methods approved by the Administrator under Sections 260.20 and 260.21.

     d. It is an oxidizer as defined in 49 CFR 173.151.

  2. A solid waste that exhibits the characteristic of ignitability must be described by the EPA hazardous waste number D001 and the hazard code for Ignitability (I).

B. **Corrosivity**

  1. **Definition.** A solid waste exhibits the characteristic of corrosivity if a representative sample of the waste has either of the following properties:

    a. It is aqueous.

    b. It has a pH less than or equal to 2 or greater than or equal to 12.5 as determined by a pH meter using either an EPA test method or an equivalent test method approved by the Administrator under the procedures set forth in Sec. 260.20 and 260.211.

The EPA test method for pH is specified as Method 5.2 in "Test Methods for the Evaluation of Solid Waste, Physical/Chemical Methods" (incorporated by reference, see Sec. 260.11) or an equivalent test method approved by the Administrator under the procedures set forth in Sec. 260.20 and 260.21.

2. It is interesting to note that EPA addresses the concept of corrosivity for liquid wastes only. While anhydrous and other dry wastes are not corrosive, given the absence of moisture or a wetting agent, they nevertheless become corrosive when exposed to environmental conditions. Although their corrosiveness cannot be measured without the presence of an ionizing medium, it is rather curious that EPA has not devised or selected a test that would specify a procedure by which the corrosivity of such materials would be considered. Still, the generator is cautioned to exercise particular care in managing those wastes that become corrosive in the presence of water.

Example. Consider a generator who decides that he wants to discard a drum of sodium hydroxide crystals (caustic soda). He sets the drum outside and waits several months before shipping it off-site. The drum becomes damaged over time. The crystals are exposed to rainwater and the resulting solution, which has a pH greater than 12.5, flows out of the container. What happens? The generator has just discharged untreated hazardous waste to the environment. He is subject to severe criminal and civil penalties. The merit of prudent and conscientious waste management is obvious.

3. A solid waste that exhibits the characteristic of corrosivity must be described by the EPA hazardous waste number D002 and the hazard code for Corrosivity (C).

C. **Reactivity**
   1. **Definition.** A solid waste exhibits the characteristic of reactivity if a representative sample of the waste has any of the following properties:
      a. It is normally unstable and readily undergoes violent change without detonating.
      b. It reacts violently with water.
      c. It forms potentially explosive mixtures with water.
      d. When mixed with water, it generates toxic gases, vapors, or fumes in a quantity sufficient to present a danger to human health or the environment.
      e. It is a cyanide- or sulfide-bearing waste which, when exposed to pH conditions between 2 and 12.5, can generate toxic gases, vapors, or fumes in a quantity sufficient to present a danger to human health or the environment.
      f. It is capable of detonation or explosive reaction if it is subjected to a strong initiating source or if heated under confinement.
      g. It is readily capable of detonation or explosive decomposition or reaction at standard temperature and pressure.
      h. It is a forbidden explosive as defined in 49 CFR 173.51, or a Class A explosive as defined in 49 CFR 173.53 or a Class B explosive as defined in 49 CFR 173.88.
   2. A solid waste that exhibits the characteristic of reactivity must be described by the EPA hazardous waste number D003 and the hazard code for Reactivity (R).
D. **EP Toxicity.** This characteristic necessarily requires the laboratory analysis of the waste in question. In general terms, the Extraction Procedure (EP) Toxicity test is a two-step process: (1) extract certain compounds from the sample; (2) analyze the extract for the EP Toxicity contaminants.
   1. **Definition.** A solid waste exhibits the characteristic of EP Toxicity if, using the test methods described in Appendix II or equivalent methods approved by the Administrator under the procedures set forth in Sections 260.20 and 260.21, the extract from a representative sample of the waste contains any of the contaminants

listed in "Table 1" [see Appendix V of this book] at a concentration equal to or greater than the respective value given in that table. Where the waste contains less than 0.5% filterable solids, the waste itself, after filtering, is considered to be the extract for the purpose of this section.

2. A solid waste that exhibits the characteristic of EP Toxicity, but is not listed as a hazardous waste in Subpart D, has the EPA Hazardous Waste Number specified in "Table 1" [see Appendix V of this book] which corresponds to the toxic contaminant causing it to be hazardous. The EPA hazard code for a waste which exhibits a characteristic of EP Toxicity is (E).

In conforming with the standards of the hazardous waste determination process, the generator must maintain a broad perspective, studying every aspect of his waste disposal objectives. One aspect that is frequently overlooked by large and small generators alike is the disposal of *empty containers*. Specific requirements for the management of residues and the containers which hold them are specified in federal and state regulations. Federal regulations, as specified in 40 CFR 261.7, are reprinted in the final appendix of this book.

A brief study of these standards yields a crucial concept: Acutely hazardous wastes residues (and the containers holding them) are more stringently controlled than other hazardous wastes. While EPA excludes from regulation various hazardous waste residues (and the containers holding them) that are *not* acutely hazardous, it specifically requires that all acutely hazardous waste residues be managed in accordance with all standards relating to the generation, transportation, treatment, storage, and disposal of hazardous waste. With respect to containers which hold a residue of an acutely hazardous waste, it should be noted that unless such a container is

- triple rinsed using a solvent capable of removing the residue, or
- cleaned by another method that has been scientifically proven to be effective,

then it is also considered an acutely hazardous waste. This is necessarily true based on the hazardous waste mixture rule stated in 40 CFR 261.3(a)(2)(iv). The entire weight of the acutely hazardous waste contaminated container must be included in the generator's calculation of hazardous waste accumulated in a calendar month. Naturally, any solvents that are used in triple rinsing this type of container must be disposed of as a hazardous waste.

Finally, the generator must realize that he must abide by DOT regulations when shipping containers off-site. Since EPA and DOT define *empty* in a different manner, the generator must still be attentive to potential violations.

Types of Waste Generated by Small Quantity Generators

TYPES OF SMALL QUANTITY GENERATORS

Before listing the types of waste generated, it is worthwhile to detail the types of generators that are likely to fit into the small generator category. Following is a partial listing of the types of organizations that are most frequently identified as small quantity generators.

Academic laboratories (secondary schools and colleges)
Airline maintenance services
Animal hospitals
Antique refinishers
Art museums and studios
Asphalt paving contractors
Automobile maintenance facilities (body and mechanical)
Biological laboratories
Building contractors
Building cleaning and maintenance services
Ceramic manufacturers

Chemical research facilities
Computer manufacturers
Copy and duplicating services
Cosmetics manufacturers
Craft supply distributors
Equipment repair shops
Exterminators (pest control)
Farm supply distributors
Fireworks manufacturers
Flooring contractors
Funeral services
Furniture manufacturers
Garden centers
Gas companies

Golf courses
Greenhouses
Hardware distributors
Hotels
Industrial gas distributors
Janitorial supply distributors
Jewelers
Laundries (dry cleaning)
Machine shops
Metal finishers
Municipal governments
Newspapers
Paint distributors
Painting contractors
Pharmaceutical laboratories

Pharmacies
Photographic supply distributors
Printers
Product formulators
Real estate companies
Research laboratories
School maintenance facilities
Service stations
Swimming pool supply distributors
Tool and die makers
Truck maintenance facilities
Truck terminals
Welding supply distributors

This list is certainly not complete, but it does show the most prominent and many of the less obvious small quantity generators. Other organizations can suddenly find themselves listed as generators due to special circumstances. A good example of this is the bank that forecloses on a commercial or residential property and discovers a quantity of hazardous materials. These materials might easily be considered hazardous waste. The bank, as owner of the facility, may be liable for any Resource Conservation and Recovery Act (RCRA) or state violations if the material is not disposed of in accordance with regulations.

WASTE STREAMS OF SMALL GENERATORS

What is the difference in waste streams between small and large generators? Assuming that small generators are not often involved in the manufacture of goods, it is certainly less likely that process wastes will be involved. Small generators are far more likely to have unused or surplus commercial products that are either listed or characteristic wastes when discarded. Some good examples are old paint, waste solvents used in stripping or refinishing operations, surplus laboratory chemicals, and old and unused pesticides.

WASTE TYPES

To simplify the discussion of waste types, we will use the following categories: solvents, acids/bases, heavy metals, pesticides, reactives, chemical reagents, and plating and heat treating wastes.

Solvents

The most common hazardous waste among small quantity generators is a broad category summarized as solvents. Most workplaces use solvents in one way or another. Solvents are used for thinning, degreasing, cleaning, and stripping, in organic synthesis. Products that remain from the redistillation of solvents or from unused paints or coatings that have been allowed to dry to an unusable form may be considered hazardous wastes if they still contain small quantities of these solvents.

While chemists for hundreds of years have searched for the "universal" solvent, water remains the most useful solvent in thousands of applications. While not regulated as hazardous in and of itself, water is present in many hazardous wastes, primarily due to its frequent use to dilute many hazardous materials as well as its use as a solvent. In this case, an aqueous solution may well be a hazardous waste, with the appropriate Environmental Protection Agency (EPA) number being that of the component that is listed as a waste. See Table I for a list of common hazardous solvents with their EPA hazard numbers.

Acids and Bases

These are the most prevalent materials commonly identified as corrosive, both by characteristic and by regulatory definition. Acids have a low pH (0–6), and bases, or alkalies, have a high pH (8–14). Acids and bases may be used by industry in such applications as stripping of metals, plating, creating solutions with metals or pesticides, treating wastewater, precipitating solids out of solution, and hundreds of other processes. Unless specifically listed, they may frequently be assigned the EPA hazardous waste number D002 due to characteristics of corrosivity. Table II has a list of common acids and bases and their EPA hazard numbers.

Table I.   Common Hazardous Solvents

Solvent	EPA Hazard Number
Acetone	F003
Benzene	F005
Butyl alcohol	F003
Carbon disulfide	F005
Carbon tetrachloride	F001
Chlorobenzene	F002
Cresols	F004
Cresylic acid	F004
Cyclohexanone	F003
o-Dichlorobenzene	F002
Ethanol	D001
2-Ethoxyethanol	F005
Ethyl acetate	F003
Ethyl benzene	F003
Ethylene dichloride	D001
Ethyl ether	F003
Isobutanol	F005
Isopropanol	D001
Kerosene	D001
Methanol	F003
Methyl ethyl ketone	F005
Methylene chloride	F001
	F002
Methyl isobutyl ketone	F003
Naphtha	D001
Nitrobenzene	F004
2-Nitropropane	F005
Petroleum solvents	D001
(flashpoint less than 140°F)	
Pyridine	F005
1,1,1-Trichloroethane	F001
	F002
1,1,2-Trichloroethane	F002
Tetrachloroethylene	
(Perchloroethylene)	F001
	F002
Toluene	F005
Trichloroethylene	F001
	F002
Trichlorofluoromethane	F002
Trichlorotrifluoroethane	
(Valclene)	F002
White spirits	D001
Xylene	F003

Table II.   Common Acids and Bases

Acid or Base	EPA Hazard Number Code
Acetic acid	D002C
Ammonium hydroxide or ammonia solution	D002C
Chromic acid	D002C
Ferric chloride	D002C
Hydrobromic acid	D002C
Hydrochloric acid or muriatic acid	D002C
Hydrofluoric acid	U134C,T
Nitric acid	D002C
Perchloric acid	D002C,R
Phosphoric acid	D002C
Potassium hydroxide	D002C
Sodium hydroxide or caustic soda	D002C
Sulfuric acid	D002C
Formic acid	U123C,T

Heavy Metals

Heavy metal is a term commonly found in the jargon of chemists and toxicologists and in environmental regulations. Heavy metals (i.e., arsenic, barium, cadmium, chromium, lead, mercury, selenium, and silver) are hazardous due to their toxic effects on various body systems. Most of these materials do not break down readily in the body, and thus can accumulate over time. Many small businesses generate wastes containing heavy metals. This may be the result of a plating operation, laboratory use, or the use of some coatings and paints. A further description of those wastes generated as a result of plating operations is under the heading entitled "Plating and Heat Treating Wastes." Metals and other waste materials that may contain small quantities of these metals have typically been considered hazardous waste according to an extraction testing procedure known as an EP toxicity test. While the specific definition of EP toxicity according to current state and federal regulations may vary, it is certain that many of the most common heavy metals will remain listed wastes (the EPA hopes to modify its toxicity definition to include a wider variety of materials that do not currently fit the EP Toxic description). Table III has a list of the heavy metals with their EPA numbers.

Pesticides

The use of herbicides, insecticides, and rodenticides is controlled by the EPA under different regulations, but the disposal of these

Table III. Common Heavy Metals

Metal	EPA Hazard Number
Arsenic	D004
Barium	D005
Cadmium	D007
Chromium	D007
Lead	D008
Mercury	D009
Selenium	D010
Silver	D011

materials comes under RCRA and state hazardous waste laws. For this reason, those businesses or institutions that discard unused quantities of these materials are considered hazardous waste generators. There are over 250 pesticides that are either currently in use or in storage. Some of these are designated as acutely hazardous and may be regulated in even very small quantities. The use of some pesticides such as DDT and Silvex has been discontinued by law due to their high toxicity. A few of the most common pesticides are listed with their EPA numbers in Table IV.

Reactives

Those materials that are generally referred to as reactive might also be best described as unstable. Reactives include those materials that react violently with air or water, as well as many strong oxidizers, shock-sensitives, heat-sensitives, and explosives. Cyanides and sulfides are considered reactive by EPA definition if they generate toxic gases or vapors at a pH between 2 and 12.5.

The very characteristic that may make one of these materials unstable is frequently the one that makes it useful in industry. The plastics industry, for example, may use a material that causes rapid polymerization to harden a plastic. A strong oxidizer may be used to change the chemical composition of a material, and many catalysts are used to speed up a reaction. Many of these materials are commonly found in laboratories, and their properties will be discussed further in the chapter on laboratories. An example of each class of reactives is listed in Table V. Nonlisted reactive materials have EPA waste number D003.

Table IV.   Common Pesticides

Pesticide	EPA Hazard Number
Aldicarb	P070
Aldrin	P004
Arsenic pentoxide	P010
Chlordane	U036
2,4 Dichlorophenoxyacetic acid	U240
DDT	U061
Dieldrin	P037
Dinoseb	P020
Endosulfan	P050
Endrin	P051
Heptachlor	P059
Hexachlorobenzene	U127
Kepone	U142
Lindane	U129
Methoxychlor	D014
Methyl parathion	P071
Nicotine	P075
Parathion	P089
Pentachlorophenol	F027
Phenylmercuric acetate	D009
2,4,5-Trichlorophenoxyacetic acid	U232
Thiram	U244
Toxaphene	P123
Warfarin	U248

Chemical Reagents

If this category of wastes defies characterization by physical state, it is clearly a single waste type in one respect; it refers to small quantities of virgin chemical, usually of high purity. Chemical reagents have a wide variety of uses, particularly in analytical or research laboratories in which substances of small quantities and high purity are required. Reagents are sold in quantities as low as ten milligrams and as high as five gallons, but most often the containers are of one pound

Table V.   Reactive Materials

Class of Reactive	Example
Organic peroxide	Benzoyl peroxide
Flammable solid	Black powder
Pyrophoric	Butyl lithium
Peroxide-forming solvent	Isopropyl ether
Water reactive	Sodium metal
Air reactive	Stannic chloride
Explosive	Lead azide

or one liter. These quantities are relatively large, considering the more specialized equipment being used today in micro- or milligram quantities in research, analytical testing, and many clinical applications. The disposal of reagents is typically managed through the practice known as labpacking, or packaging reagents in containers specified by the U.S. Department of Transportation, usually of steel or fiber, and 30 or 55 gallons in size. Additional information on labpacks will be supplied in the chapter on laboratories.

Plating and Heat Treating Wastes

Many small businesses perform plating operations that may generate hazardous waste. The most common of these wastes are those from electroplating, anodizing, and salt bath heat treatment. These solutions frequently contain cyanides, since their physical and chemical properties make them ideal for plating, particularly chrome, silver, copper, and brass. Other plating and heat treatment methods involve much more expensive procedures. Until alternative procedures can be developed that provide efficiency similar to those utilizing cyanides, we can expect that these wastes will continue to be produced by many small generators.

There are a number of ways to treat cyanides to render them less harmful. Most of these involve the use of either chlorine gas or a chlorinated oxidizer such as sodium hypochlorite, to convert the cyanides to cyanates. Further treatment with sodium hydroxide can convert these cyanates to wastes that are relatively harmless.

Most of the plating or heat treating wastes have EPA waste numbers F006 through F012, depending on the specific process. These wastes are listed in 40 CFR 261.31 as hazardous wastes from nonspecific sources.

4

Generator Requirements

In terms of the quantity and types of wastes produced or accumulated in a calendar month, a hazardous waste generator will fit into one of three categories as defined by regulation. These categories are

- conditionally exempt generators
- small quantity generators (SQGs)
- large generators

Chapter 2 describes these in detail. Each category is subject to regulatory standards to varying degrees, with the regulations becoming increasingly more stringent as each threshold of quantity of waste generated is crossed.

In determining the regulatory requirements that are applicable, the generator must first determine whether his state (or territory) has been granted final authorization to administer the Resource Conservation and Recovery Act (RCRA) program. As discussed in Chapter 2, this is a critical step due to the significant differences between certain state programs and the federal program. Such differences might be evidenced in several ways:

- the scope and range of the types of wastes that are specifically excluded or included in the definition of solid and hazardous waste;

- the threshold quantity for hazardous waste that is either generated or stored in a calendar month (see Appendix I for state listings); or
- the applicability of standards to various generators' situations.

In essence, these differences can result in significant disparities between federal and state rules regarding the applicability of requirements such as notification, length of accumulation time, emergency preparedness, and manifesting. As a result, the first step in complying with generator requirements is to understand whether federal or state rules apply. Appendix I of this handbook delineates those states that currently have primacy.

The generator is reminded also, that in many cases, states which are authorized to administer the RCRA program are not always up to pace with recently revised, more stringent federal standards. In cases where such a state's requirements are less stringent than the federal requirements, the generator *must* abide by the federal regulations. For example, prior to the Spring of 1987, the Commonwealth of Pennsylvania was granted final authorization to administer its own hazardous waste regulations. Pennsylvania's current regulations for SQGs of hazardous waste are less stringent than current federal regulations. These generators, therefore, must comply with federal standards for small quantity generators but must work within Pennsylvania's code for other waste management rules. Eventually, all states must incorporate the Environmental Protection Agency's (EPA's) standards into their programs according to a compliance schedule that meets criteria established by EPA. Thus, a generator might be subject to federal regulation even in a state which has primacy.

I. Conditionally exempt generators

   A. Required actions. Of the three types of generators of hazardous waste, conditionally exempt generators are subject to the fewest regulatory controls. In essence, these controls require three actions:

      1. Comply with 40 CFR 262.11 concerning hazardous waste determination

      2. Send any hazardous waste to a hazardous waste facility, legitimate recycling facility, or any other state-authorized facility that is authorized to manage industrial or municipal wastes

      3. Never accumulate in a calendar month more than any of the following quantities:

   a. 1 kg of acutely hazardous waste

   b. 100 kg of a spill residue generated from the cleanup of an acutely hazardous waste

   c. 1000 kg of hazardous waste

B. Maintaining conditionally exempt status. If a conditionally exempt generator exceeds any of these threshold quantities, he must manage his hazardous waste in accordance with other generator standards, namely, those relating to either small or large quantity generators. Naturally, the quantity and types of wastes that are stored or accumulated determine whether the small or large quantity generator standards apply.

In terms of financial liability, it may not be advisable for a generator to remove hazardous waste frequently for the sole purpose of maintaining conditionally exempt status. This decision must take a number of factors into consideration, including short-term cost. In terms of potential legal liabilities, public scrutiny, and the effort that is required to institute and maintain a hazardous waste management program, it is always in the best interest of a generator to maintain his conditionally exempt status. This is not to say, however, that a generator should intentionally misrepresent the facts concerning the types and quantities of waste generated or stored. Every citizen has the responsibility to comply with the regulations that are intended to protect human health and the environment. A generator can preserve his conditionally exempt status in a number of ways, including:

1. Reducing the quantity and toxicity of waste
2. Using less toxic products in the normal course of operations
3. Avoiding excessive inventories of toxic chemicals, thereby reducing potential spill and disposal problems
4. Providing timely treatment or off-site disposal of wastes before they are accumulated in large quantities

There is a significant difference between the requirements for conditionally exempt generators and small quantity generators of hazardous waste. The requirements for SQGs are described below.

II. Small quantity generators
A. Notification
1. **Definition.** A generator is required to obtain an EPA identification number **(EPA ID#)** before he treats, stores, disposes of, transports, or offers hazardous waste for transportation. The process of obtaining this number is typically referred to as **notification**. It is important to recognize that this notification is not considered an application, license, or permit.
2. **Process.** The process of notification is relatively straightforward. Once a generator has determined that he generates the quantities and types of waste that make him a SQG, he must follow several steps:
   a. Contact the state hazardous waste management agency or EPA and request a copy of **EPA Form 8700–12, "Notification of Hazardous Waste Activity."** Fairly detailed instructions are provided with this simple, two-page form (Figures 1 and 2).
   b. Complete the form. Take particular notice that the form distinguishes between the **mailing address** and the **site address** for operation relating to hazardous waste activity. This distinction is important because EPA ID#s are location-specific. That is, if the waste generator has multiple sites where he conducts business which result in the generation of regulated quantities of hazardous waste, he must complete a form for each location. It is conceivable that each site's mailing address will be identical. For example, consider a dry cleaner who manages 10 different locations from one office. A form should be completed for each site. Although each form will show a unique site address, they would all have the same mailing address. Each plant will ultimately be assigned a unique EPA ID#.
3. **Updating the EPA**
   a. Since EPA ID#s are location-specific, the generator is required to notify EPA or the appropriate state agency if he moves his business to a **different location.**
   b. Furthermore, especially **if he sells his business** to another individual or company, the generator

Please print or type with ELITE type *(12 characters/inch)* in the unshaded areas only.

Form Approved OMB No. 2000-0098
GSA No. 0246-EPA-OT Expiration Date 12/31/86

⊕EPA

U.S. ENVIRONMENTAL PROTECTION AGENCY
NOTIFICATION OF HAZARDOUS WASTE ACTIVITY

INSTRUCTIONS: If you received a preprinted label, affix it in the space at left. If any of the information on the label is incorrect, draw a line through it and supply the correct information in the appropriate section below. If the label is complete and correct, leave Items I, II, and III below blank. If you did not receive a preprinted label, complete all items. "Installation" means a single site where hazardous waste is generated, treated, stored and/or disposed of, or a transporter's principal place of business. Please refer to the INSTRUCTIONS FOR FILING NOTIFICATION before completing this form. The information requested herein is required by law *(Section 3010 of the Resource Conservation and Recovery Act).*

INSTALLATION'S EPA I.D. NO.

I. NAME OF INSTALLATION

II. INSTALLATION MAILING ADDRESS

III. LOCATION OF INSTALLATION

PLEASE PLACE LABEL IN THIS SPACE

FOR OFFICIAL USE ONLY

COMMENTS

C

INSTALLATION'S EPA I.D. NUMBER | APPROVED | DATE RECEIVED *(yr., mo., & day)*

F

I. NAME OF INSTALLATION

II. INSTALLATION MAILING ADDRESS

STREET OR P.O. BOX

3

CITY OR TOWN | ST. | ZIP CODE

4

III. LOCATION OF INSTALLATION

STREET OR ROUTE NUMBER

5

CITY OR TOWN | ST. | ZIP CODE

6

IV. INSTALLATION CONTACT

NAME AND TITLE *(last, first, & job title)* | PHONE NO. *(area code & no.)*

2

V. OWNERSHIP

A. NAME OF INSTALLATION'S LEGAL OWNER

8

B. TYPE OF OWNERSHIP *(enter the appropriate letter into box)*

F = FEDERAL
M = NON-FEDERAL

VI. TYPE OF HAZARDOUS WASTE ACTIVITY *(enter "X" in the appropriate box(es))*

☐ A. GENERATION   ☐ B. TRANSPORTATION *(complete item VII)*
☐ C. TREAT/STORE/DISPOSE   ☐ D. UNDERGROUND INJECTION

VII. MODE OF TRANSPORTATION *(transporters only — enter "X" in the appropriate box(es))*

☐ A. AIR   ☐ B. RAIL   ☐ C. HIGHWAY   ☐ D. WATER   ☐ E. OTHER *(specify)*:

VIII. FIRST OR SUBSEQUENT NOTIFICATION

Mark "X" in the appropriate box to indicate whether this is your installation's first notification of hazardous waste activity or a subsequent notification. If this is not your first notification, enter your installation's EPA I.D. Number in the space provided below.

☐ A. FIRST NOTIFICATION   ☐ B. SUBSEQUENT NOTIFICATION *(complete item C)*

C. INSTALLATION'S EPA I.D. NO.

IX. DESCRIPTION OF HAZARDOUS WASTES

Please go to the reverse of this form and provide the requested information.

EPA Form 8700-12 (6-85)

CONTINUE ON REVERSE

Figure 1. EPA Form 8700–12, "Notification of Hazardous Waste Activity." (front)

I.D. – FOR OFFICIAL USE ONLY

W

IX. DESCRIPTION OF HAZARDOUS WASTES *(continued from front)*

A. HAZARDOUS WASTES FROM NON–SPECIFIC SOURCES. Enter the four–digit number from 40 CFR Part 261.31 for each listed hazardous waste from non–specific sources your installation handles. Use additional sheets if necessary.

1	2	3	4	5	6
7	8	9	10	11	12

B. HAZARDOUS WASTES FROM SPECIFIC SOURCES. Enter the four–digit number from 40 CFR Part 261.32 for each listed hazardous waste from specific industrial sources your installation handles. Use additional sheets if necessary.

13	14	15	16	17	18
19	20	21	22	23	24
25	26	27	28	29	30

C. COMMERCIAL CHEMICAL PRODUCT HAZARDOUS WASTES. Enter the four–digit number from 40 CFR Part 261.33 for each chemical substance your installation handles which may be a hazardous waste. Use additional sheets if necessary.

31	32	33	34	35	36
37	38	39	40	41	42
43	44	45	46	47	48

D. LISTED INFECTIOUS WASTES. Enter the four–digit number from 40 CFR Part 261.34 for each listed hazardous waste from hospitals, veterinary hospitals, medical and research laboratories your installation handles. Use additional sheets if necessary.

49	50	51	52	53	54

E. CHARACTERISTICS OF NON–LISTED HAZARDOUS WASTES. Mark "X" in the boxes corresponding to the characteristics of non–listed hazardous wastes your installation handles. *(See 40 CFR Parts 261.21 – 261.24.)*

☐ 1. IGNITABLE (D001) ☐ 2. CORROSIVE (D002) ☐ 3. REACTIVE (D003) ☐ 4. TOXIC (D000)

X. CERTIFICATION

I certify under penalty of law that I have personally examined and am familiar with the information submitted in this and all attached documents, and that based on my inquiry of those individuals immediately responsible for obtaining the information, I believe that the submitted information is true, accurate, and complete. I am aware that there are significant penalties for submitting false information, including the possibility of fine and imprisonment.

SIGNATURE	NAME & OFFICIAL TITLE *(type or print)*	DATE SIGNED

EPA Form 8700-12 (6-85) REVERSE

DETACH

Figure 2. EPA Form 8700–12, "Notification of Hazardous Waste Activity." (reverse)

should always provide a mechanism by which he will be assured that the buyer will submit an amended Form 8700–12. This will minimize the seller's exposure to complications in the event that the buyer acts in a manner that instigates an enforcement action by an appropriate regulatory agency. Unless the form is amended to show that the location is under different ownership, the seller is likely to get caught in the middle between the EPA and the new owner in the event of an enforcement action.

c. If the generator decides to **cease all operations** that produce hazardous waste, he should contact his state agency or regional EPA office to determine the best method of documenting the termination of his waste management activities.

4. **Eligible recipients of waste.** In addition to obtaining an EPA ID#, the generator is also required to offer his waste only to entities which have obtained ID#s for their own activities. This requirement is intended to ensure that all hazardous wastes are managed within the scope of RCRA at the federal and state levels. EPA maintains a database of EPA ID#s and uses this information for a variety of purposes ranging from the gathering of information for future policies to enforcement actions.

B. **The hazardous waste manifest.** The hazardous waste manifest is an extremely important document for the off-site shipment of hazardous waste. For practical purposes, the discussion of this document's use must be limited here.

1. **When to use.** Small quantity generators must always use the uniform hazardous waste manifest, except under the following conditions:

a. The waste is reclaimed under a contractual agreement in which:

(1) The type of waste and frequency of shipments are specified in the agreement;

(2) The vehicle used to transport the waste to the recycling facility and to deliver reclaimed material back to the generator is owned and operated by the reclaimer of the waste; and

b. The generator maintains a copy of the reclamation agreement in his files for a period of at least three years after termination or expiration of the agreement. [40 CFR 262.20]

2. Types

a. Federal. The **uniform manifest** can be considered a document that serves informational requirements for two levels of government — federal and state. These requirements are represented on the form by two conventions: (1) shading and (2) alpha or numeric characters. Notice in the sample reprinted in this chapter (Figure 3) that certain items are assigned numerals (e.g., "11" precedes "U.S. DOT [Department of Transportation] Description") while other items are preceded by letters (e.g., "H" is associated with "Facility's Phone").

The **nonshaded numbered items** on the manifest are those which are required to be completed on every manifest, regardless of the state to or from which hazardous waste is being sent or received. The **shaded, alphabetized items** on the manifest are provided for state-specific requirements. These informational items cause significant confusion in the regulated community due largely to the fact that there is some inconsistency between states for completion of these items. For example, New Jersey requires the completion of Item B on the manifest, whereas neighboring state Pennsylvania does not.

b. State-specific. States with final authorization to administer regulations are permitted to print their **own versions of the uniform manifest.** Although these states must preserve the format and informational headings of the national uniform manifest, they are allowed to require that additional information be provided in the shaded portions of the manifest. This, coupled with the fact that some states regulate more types of waste than others, means proper completion of the manifest can be a difficult task.

3. Acquisition. Recognizing the confusion that is caused

Part 262, Appendix **40 CFR Ch. I (7-1-86 Edition)**

Please print or type. *(Form designed for use on elite (12-pitch) typewriter.)* *Form Approved. OMB No. 2000-0404. Expires 7-31-86*

UNIFORM HAZARDOUS WASTE MANIFEST

1. Generator's US EPA ID No. Manifest Document No.
2. Page 1 of
Information in the shaded areas is not required by Federal law.

GENERATOR

3. Generator's Name and Mailing Address
4. Generator's Phone ( )
5. Transporter 1 Company Name
6. US EPA ID Number
7. Transporter 2 Company Name
8. US EPA ID Number
9. Designated Facility Name and Site Address
10. US EPA ID Number

A. State Manifest Document Number
B. State Generator's ID
C. State Transporter's ID
D. Transporter's Phone
E. State Transporter's ID
F. Transporter's Phone
G. State Facility's ID
H. Facility's Phone

11. US DOT Description *(Including Proper Shipping Name, Hazard Class, and ID Number)*	12. Containers No.	12. Containers Type	13. Total Quantity	14. Unit Wt/Vol	I. Waste No.
a.					
b.					
c.					
d.					

J. Additional Descriptions for Materials Listed Above

K. Handling Codes for Wastes Listed Above

15. Special Handling Instructions and Additional Information

16. **GENERATOR'S CERTIFICATION:** I hereby declare that the contents of this consignment are fully and accurately described above by proper shipping name and are classified, packed, marked, and labeled, and are in all respects in proper condition for transport by highway according to applicable international and national government regulations.

Unless I am a small quantity generator who has been exempted by statute or regulation from the duty to make a waste minimization certification under Section 3002(b) of RCRA, I also certify that I have a program in place to reduce the volume and toxicity of waste generated to the degree I have determined to be economically practicable and I have selected the method of treatment, storage, or disposal currently available to me which minimizes the present and future threat to human health and the environment.

Printed/Typed Name | Signature | *Month Day Year*

TRANSPORTER

17. Transporter 1 Acknowledgement of Receipt of Materials

Printed/Typed Name | Signature | *Month Day Year*

18. Transporter 2 Acknowledgement of Receipt of Materials

Printed/Typed Name | Signature | *Month Day Year*

FACILITY

19. Discrepancy Indication Space

20. Facility Owner or Operator: Certification of receipt of hazardous materials covered by this manifest except as noted in Item 19.

Printed/Typed Name | Signature | *Month Day Year*

EPA Form 8700-22 (Rev. 4-85) Previous edition is obsolete.

Figure 3. EPA Form 8700–22, "Uniform Hazardous Waste Manifest."

by state-specific uniform manifests and acknowledging that each state can administer a RCRA program that is more stringent than the federal program, EPA promulgated a regulation with respect to the acquisition of manifests by generators of hazardous waste. By establishing such a regulation, EPA provides a control, so that state-authorized programs can gather the type of information that is provided by a properly completed state uniform manifest. As given in 40 CFR 262.21, this regulation states:

a. "If the State to which the shipment is manifested **(consignment State)** supplies the manifest and requires its use, then the generator must use that manifest.

b. If the consignment State does not supply the manifest, but the State in which the generator is located **(generator State)** supplies the manifest and requires its use, then the generator must use that State's manifest.

c. If neither the generator State nor the consignment State supplies the manifest, then the generator may obtain the manifest from any source."

4. **Number of required copies.** In addition to complying with the requirements concerning the acquisition and completion of the manifest, the SGQ should also understand the requirement relating to the number of required copies. As stated in 40 CFR 262.22:

The manifest consists of at least the number of copies which will provide the generator, each transporter and the owner or operator of the designated facility with one copy each for their records and another copy to be returned to the generator.

5. **Retention.** Federal regulations [40 CFR 262.40 (a)] require that the generator maintain a copy of the manifest for three years. Some states may require longer retention times.

6. **Submission of the completed manifest.** With respect to the number of copies of the manifest, the biggest difference in the regulations for SQGs is that SQGs are generally not required to submit copies of their manifests to the state. Whereas large quantity generators must use an eight-part form, the SQG usually needs only a four-part form. At least one state, New York, supplies a four-part manifest for use by SQGs.

It is recommended that unless specifically required to do so, SQGs should not submit manifest copies to the state. Submitting manifests when it is not required may not only impede efficient waste management administration by the state agency, but it may also result in enforcement action if the manifest is noted by the agency to have been improperly completed. Such simple mistakes as using a wrong abbreviation or not submitting clearly legible copies may result in fines to the generator.

7. **Proper completion.** Proper completion of the hazardous waste manifest is a fairly simple matter if the generator takes the time to carefully read the instructions provided on the back side of the manifest document. The appendix of 40 CFR Part 262 (Appendix VI of this book) also offers clear instructions for proper completion of the form. A few comments regarding some of the required information are worth discussing.
   a. **Unshaded areas**
      (1) **Item 1.** This item actually consists of two parts: the generator's EPA identification number and the manifest number. It is important that the generator maintain a log of manifest numbers that he uses, since each number gives each waste item or shipment a unique character. The log will assist the generator in tracking each number that he has assigned a manifest, and it will reduce the probability of confusing one waste shipment with another. The generator need not be concerned with assigning the same manifest number that thousands of other generators might use, because each generator's unique EPA ID# precedes the manifest document number. In essence, EPA identifies each manifest by a 17-character code.
      (2) **Item 2.** "Page 1" specified in this item should simply be viewed as the title or name of this box. The generator should always be sure to fill in the blank spaces to the left and the right of the word **of** in this box. For example, when using a manifest without a continuation sheet

for additional items, this item should show "1 of 1."

(3) **Item 11.** Always be sure to enter the U.S. DOT shipping description in accordance with 49 CFR Part 173. Be especially careful to note "RQ" whenever reportable quantities are involved.

(4) **Item 13.** Always manifest the **actual** quantity of waste. It is a sloppy recordkeeping practice to simply assume that each container is full. This might also lead someone to the assumption that a container either leaked or was partially emptied during transit. Also, some disposal facilities might bill a higher amount for disposal of material that is not actually in a container.

(5) **Item 14.** Choose an appropriate unit of measure. Many people commit the error of marking "G" (gallons) for nonliquid waste.

(6) **Item 15.** This space provides an excellent opportunity for noting various types of information, including:

(a) An alternate facility. This could save the generator from having to accept waste returned by the primary facility which could not be handled for one reason or another.

(b) Spill cleanup precautions in the event of a discharge

(c) Treatment, storage and disposal (TSD) facility authorization numbers relating to acceptance of the waste

(d) Billing information

(e) Listing hazardous materials and other non-RCRA regulated wastes

b. **Shaded areas**

(1) **Item A.** Many states supply their own manifest document numbers for internal purposes. (Whether such a number is supplied or not, the generator should always indicate his own manifest number in Item 1, an unshaded area.)

(2) **Item B.** Some states require that the generator

note the location where the waste is generated, as opposed to the generator's mailing address. In cases where the mailing address and site address are identical, it is usually advisable to indicate "Same" in Item B. Failure to do so may be construed by some state regulatory agencies as a violation.

(3) **Item C.** This information typically refers to the waste transporter's state hazardous waste hauler permit or truck number. As mentioned in other chapters of this book, some states require hazardous waste transporters to apply for special waste hauling permits. If a transporter meets certain criteria, it is assigned a special number. It is this number which should be entered in Item C. This number is usually not identical to the transporter's EPA ID#, though it might be similar.

(4) **Item I.** The EPA hazardous waste number or the state's waste number for a waste stream must be entered here. In cases where a waste is described by multiple numbers, all numbers should be entered here. For example, a solvent mixture containing greater than 10% methylene chloride, acetone, and toluene would be described as F001, F003, and F005.

This item causes substantial problems when labpacks are manifested. While EPA and many states provide for the use of continuation sheets for the manifesting of labpacks, some states do not. This can cause serious space problems for documenting the numerous listed wastes that are contained in a given labpack. In such a situation, the generator should contact his state agency for advice.

(5) **Item J.** Usually the information requested in Item J relates to a waste's physical state and hazard code. Some states, however, require more detailed information about the waste. The instructions on the manifest should indicate the required information.

(6) Item K. Some states require the generator to specify the intended treatment/disposal method for each waste stream indicated in Item 11. Codes for treatment/disposal methods are referenced on the back of the manifest.

C. **Pretransport requirements.** Referenced in 40 CFR Part 262 Subpart C, these requirements have several objectives:

- to ensure generator **compliance with U.S. DOT requirements** relating to the shipment and transportation of hazardous waste
- to establish **time constraints** regarding the type and quantity of hazardous waste that can be stored on-site
- to require that measures be taken to **prevent accidents**
- to establish basic standards for **responding to fire, explosion, or the discharge of hazardous waste**

1. **Activities required by U.S. DOT.** This objective is a perfect example of the concept of "incorporation by reference." That is, EPA has simply adopted standards established by U.S. DOT by referencing particular part numbers of 49 CFR. Hence, hazardous waste generators, both large and small, must comply with the details of the hazardous material regulations established by U.S. DOT.

   a. **Packaging.** U.S. DOT references the types of packages that are required for various types of materials in the Hazardous Materials Table referenced in 49 CFR 172.101 in Column 5. To determine the type of package that is permitted for a particular material, the generator simply refers to that section of 49 CFR which is referenced in the table.

   Example. A generator has 20 gal of mixed solvents including methylene chloride, toluene, and methyl ethyl ketone. He has followed all of the steps of the hazardous waste determination process and knows

   - that the waste mixture has a flash point of 90°F;
   - that the EPA waste type is F001, F005; and
   - that the EPA hazard code is (I,T).

   In order to effectively use the Hazardous Materials Table in 49 CFR 172.101, the generator must determine whether this material fits into any of the **DOT** definitions of a hazardous material or waste. (He may refer to Appendix VII of this book or to the definitions provided throughout 49 CFR Part 173.) The generator discovers three important facts:

- he must assign the mixture a general (nonspecific) shipping name since it is composed of more than one chemical;
- that the mixture meets the definition of "Flammable Liquid" per 49 CFR 173.115; and
- the only suitable name for this mixture is "Waste Flammable Liquid, n.o.s."

Next, the generator reads the references in column 5b for the shipping name "Waste Flammable Liquid, n.o.s." and opens 49 CFR to Part 173.119. Here, he discovers a number of container types that he is permitted to use and selects an appropriate package.

b. **Labeling.** The labeling requirements for packages [49 CFR Subpart E] basically require that the shipper provide information concerning his name, address, proper shipping name, DOT hazard class, EPA hazard class, and waste type number and DOT identification number. The details of the labeling requirements become quite complex for certain types of materials. A reliable transporter or waste management company can offer assistance in ensuring compliance in this area. In fact, many transporters will provide and complete the labels for the generator.

    With the exception of studying the regulations in painstaking detail, perhaps the best way to understand the labeling regulations is to carefully observe the type of information that was provided on the label with the virgin materials that were received at the site. Major chemical manufacturing and distribution companies generally comply with

DOT regulation in exemplary fashion. In formulating the proper labeling information, however, always remember to describe the waste material as "Waste."

c. **Marking.** Marking differs from labeling in that it describes the physical attributes of a marker (or label) affixed to a container (Figure 4), whereas labeling refers to the information required on the marking. Specified in 40 CFR 262.32 and 49 CFR Part 172, perhaps the most important aspect of this regulation is that a hazardous waste container must be provided with a special marking, as follows:

HAZARDOUS WASTE—Federal Law Prohibits Improper Disposal. If found, contact the nearest police or public safety authority or the U.S. Environmental Protection Agency.

Generator's name and address_____

_____

Manifest Document Number_____

It is important to note that many states require additional information on the marking. The generator should always check the generator requirements of his state's hazardous waste regulations to ensure compliance.

d. **Placarding.** EPA also requires that the generator comply with the requirements of 49 CFR Part 172 Subpart F, which relates to the placarding of vehicles for transporting hazardous waste. The purpose of placarding is to assist first responders to an environmental incident in identifying hazards associated with the vehicle. Fortunately, placarding requirements are generally straightforward, as presented in Tables 1 and 2 of 49 CFR 172.504.

Again, nearly every reliable transporter will supply placards (Figure 5) and properly use them on his vehicle. The small quantity generator should be extremely leery of using a transporter who does not have placards and the knowledge of placarding re-

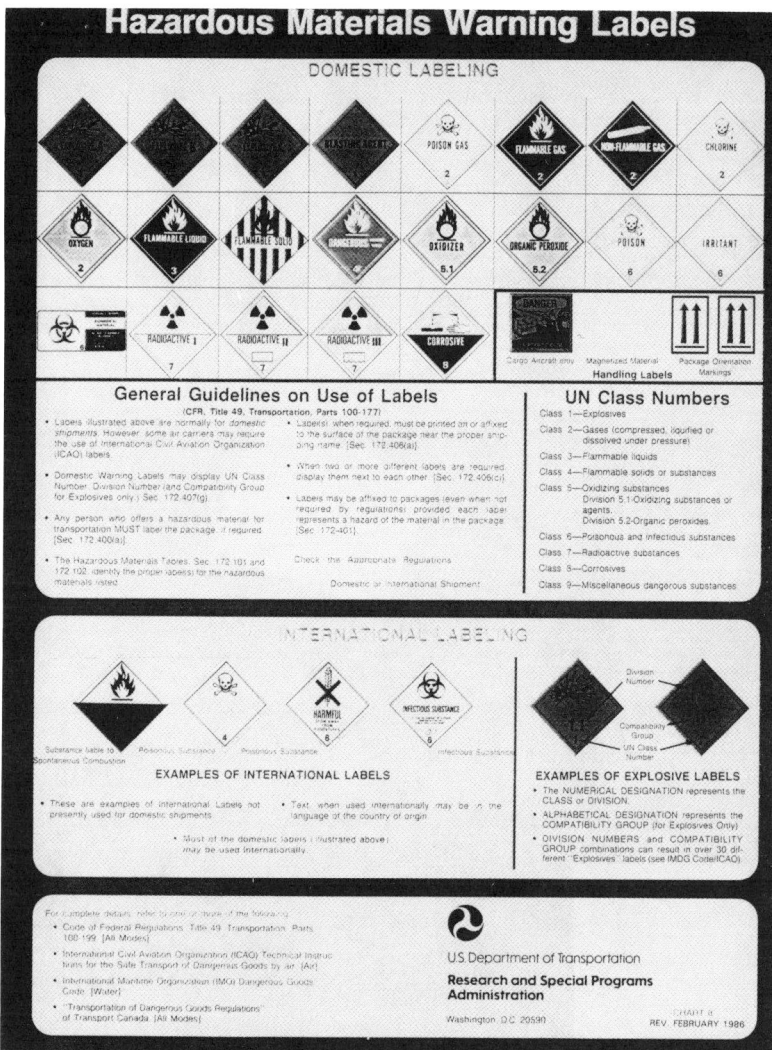

Figure 4.   Hazardous materials warning labels.

quirements. The generator should, however, verify that the vehicle is properly placarded before it leaves his site, since supplying them is actually his responsibility.

2. **Accumulation time.** The second basic objective of the pretransport requirements is to place time constraints

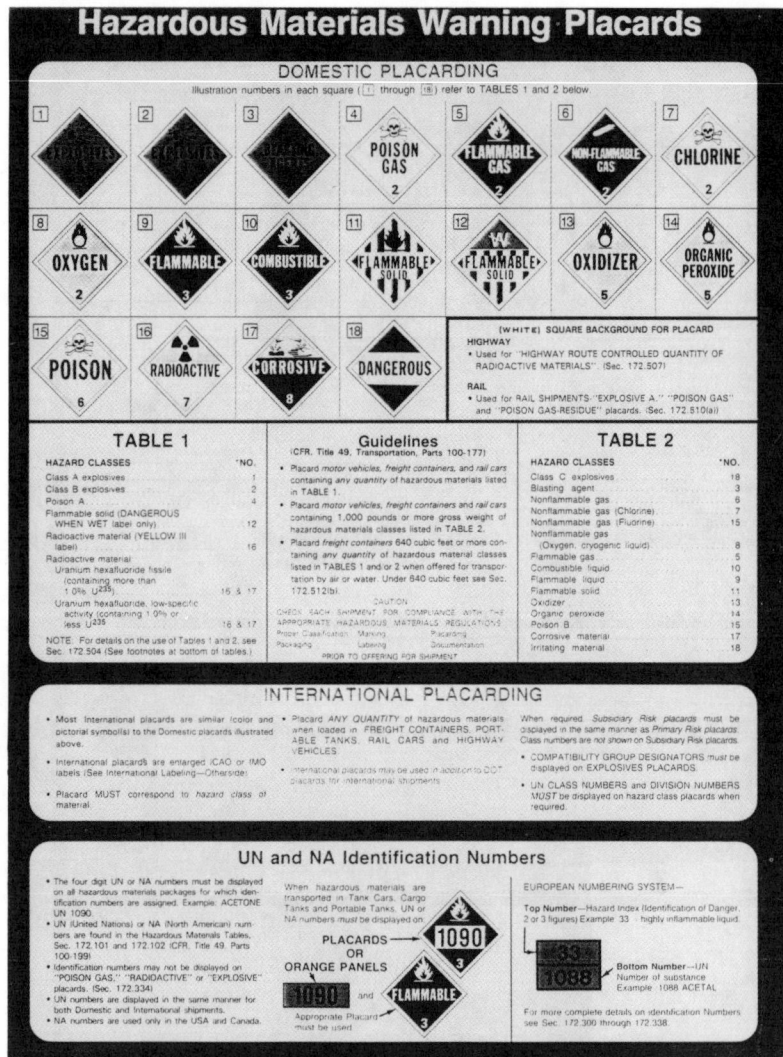

Figure 5. Hazardous materials warning placards.

on the quantity and type of waste that a generator may store on-site without needing a permit to engage in storage activities. These constraints are specified in 40 CFR 262.34 as **accumulation time.** This section distinguishes between small and large quantity generators.

The following discussion deals specifically with small quantity generators.

a. **Safe containers.** Prior to commencing storage of any waste type for any period of time, the first factor that the SQG must consider in storing hazardous waste is the waste's **compatibility** with its intended container. If he complies with U.S. DOT packaging requirements, the generator will always ensure waste/container compatibility. That is, there will be no dangerous evolution of heat or gas, nor will there be a reaction between the container and its contents that will result in the deterioration of the container's structure or integrity.

Naturally, any type of container that is selected for the storage of hazardous waste should be **free of corrosion, dents, damaged closures,** or any other condition that would impede its ability to provide for the safe containment of its contents.

b. **Safe management of containers.** The generator must manage his waste containers in a manner that will not result — potentially or actually — in the release of hazardous waste. The generator should always ensure that storage containers are **kept closed** at all times, except when waste is being added to or removed from them. Similarly, the generator should provide a location for waste containers in areas where there is **no chance for damage** due to operations. An important aspect of ensuring proper container management is to establish **regular inspections** of containers. Such inspections should be well documented and should consider such factors as leakage, closure, rust or deterioration, location, the condition of the floor or area in which the containers are stored, and potential fire, health, and safety hazards associated with the storage of these containers. Containers should **never** be stored in an area where wastes can flow into sanitary or stormwater drains.

c. **Ignitables and reactives.** An additional requirement for containers relates to those that contain ignitable and/or reactive wastes. (Remember, a waste can be designated as ignitable or reactive either according to the hazard class associated with its listing or according to its property of exhibiting such a hazardous waste characteristic.) **Generators of ignitable or reactive wastes are required to store such wastes at least 50 feet from the property of others.** Several states are even more restrictive about the required distance. Further, such wastes must be stored according to the recommendations of the National Fire Protection Association (NFPA). An example of NFPA's requirements for ignitable or reactives storage, e.g., 55-gallon drums, is to store drums in the following configuration: (1) not more than two wide with no stated limit on aisle length, with at least 5 feet of clearance completely around the aisle, and (2) stacked no more than two drums high.

These management practices are referenced in 40 CFR 262.34 and are explicitly stated in 40 CFR 265.170–177. Flammable containers of 5 gallons and larger should also be grounded to prevent static charge buildup during transfer operations. The SQG should always check the requirements of his state to ensure compliance with standards that may be more stringent.

d. **Tracking number of days waste is stored.** Once the generator has determined suitable container types, storage locations, and safe operating practices for managing wastes on-site, he must track the type and quantity of waste and the number of days that he stores his waste. According to 40 CFR 262.34 (d), "a generator who generates greater than 100 kg (220 lb) but less than 1000 kg (2200 lb) of hazardous waste in a calendar month may accumulate hazardous waste on-site for 180 days or less without a permit (or 270 days if waste must be transported to a facility over 200 miles away) or without having interim status as a TSD facility provided that:

(1) The quantity of waste accumulated on-site

never exceeds 6000 kg (approximately 13,200 lb);

(2) The generator manages his waste in compliance with Subpart I or J of Part 265 [as discussed previously] and marks each waste container with the words 'Hazardous Waste';

(3) The generator clearly marks on the container the date on which accumulation begins [waste first added to the container];

(4) The generator complies with requirements relating to accident prevention and contingency planning [discussed below]."

e. **Tracking quantity of waste stored.** The generator is strongly cautioned to carefully monitor the quantity accumulated of acutely hazardous waste and spill residue contaminated with acutely hazardous waste. Storing more than 1 kg or 100 kg of either, respectively, for more than 90 days will necessarily mean that he is classed as a large quantity generator. (Recall the various definitions of a generator shown in Chapter 2.) It is also important to note that:

(1) Some authorized states do not recognize small quantity generators and therefore do not permit generators to store wastes for more than 90 days without a permit;

(2) Some authorized states recognize small quantity generators, but at much lower monthly rates of hazardous waste generation, and permit lower total quantities of waste to be stored on-site;

(3) Some authorized states have adopted the federal standards relating to the amount of time that a generator can store and accumulate various types and quantities of waste. Appendix I summarizes the different limits on hazardous waste accumulation for each state, listing those which are identical to the federal standards and how the others differ.

3. **Preparedness and prevention.** As previously stated, the third basic objective of establishing pretransport re-

quirements for hazardous waste generators is to ensure that adequate measures are taken to prevent accidents, which can result in the release of a hazardous waste from its container. Per 40 CFR 262.34 (d)(3), the small quantity generator is permitted to store wastes for periods up to 270 days without a permit if, among other things, he complies with the requirements of Part 265 Subpart C, namely, Preparedness and Prevention. (Some of the requirements of Part 265 Subpart C are reprinted in Appendix VIII of this book.) These standards specifically require:

a. "The maintenance and availability of communications and firefighting equipment;

b. Ensuring adequate aisle space to permit access of personnel and equipment to any area of the facility;

c. Attempting to make arrangements with State environmental authorities and local fire, police, and medical authorities and contractors concerning emergency planning."

4. **Contingency procedures.** Finally, the fourth basic objective of establishing Subpart C of Part 262 is to ensure that generators comply with certain procedures when responding to a spill or accident on their site. The actual regulations which specify contingency procedures are referenced in 40 CFR 262.34 (d) and are reprinted in Appendix IX of this book.

The authors of EPA publication 530-SW-86–019 summarize the importance of planning for emergencies (page 15): "A contingency plan is a plan that attempts to look ahead and prepare for any accidents that could possibly occur. It can be thought of as a set of answers to a series of 'what if' questions. For example: 'What if there is a fire in the area where hazardous waste is stored?' or 'What if I have a spill of hazardous waste or one of my containers leaks?' Emergency procedures are the steps you should follow if you have an emergency, that is, if one of the 'contingencies' or 'what ifs' occurs. While a specific written contingency plan is not required, it may be a good idea to make a list of these questions and answer them on paper. This also may be helpful in informing your employees about their responsibilities in the event of an emergency.

"If you have an emergency in your plant:

a. In the event of a **fire**, call the fire department or attempt to extinguish it using the appropriate type of fire extinguisher.

b. In the event of a **spill,** contain the flow of hazardous waste to the extent possible and notify the National Response Center. The Center operates a 24-hour toll free number: 800–424–8802, or in Washington, DC: 426–2675. As soon as possible, clean up the hazardous waste and any contaminated materials or soil.

c. In the event of a **fire, explosion, or other release,** immediately notify the National Response Center as required by Superfund regulations. (Superfund is the law that deals with the cleanup of spills and leaks of hazardous waste at abandoned hazardous waste sites.)

Emergency phone numbers and locations of emergency equipment must be posted near telephones and all employees must know proper waste handling and emergency procedures. You must appoint an employee to act as **emergency coordinator** to ensure that emergency procedures are carried out in the event an emergency arises. The responsibilities of the emergency coordinator are generally that he/she be available 24 hours a day (at the facility or by phone) and know whom to contact and what steps to follow in an emergency. For most small businesses, the owner or operator may already perform these functions. Thus, it is not intended nor is it likely that you will need to hire a new employee to fill this role.

It is important to avoid potential risks in this area. If you have a serious emergency and you have to call your local fire department or you have a spill that extends outside your plant or that could reach surface waters, IMMEDIATELY CALL THE NATIONAL RESPONSE CENTER (800–424–8802) AND GIVE THEM THE INFORMATION THEY ASK FOR. If you didn't need to call, they will tell you so. BUT ANYONE WHO WAS

SUPPOSED TO CALL AND DOES NOT IS SUBJECT TO A $10,000 FINE, A YEAR IN JAIL, OR BOTH. An owner or manager of a business who fails to report a release also may have to pay for the entire cost of repairing any damage, even if the facility was not the single or the main cause of the damage."

D. **Recordkeeping and reporting.** The small quantity generator must comply with certain recordkeeping requirements concerning manifests and records pertinent to hazardous waste determination. Unlike a large generator, he is not required to submit biennial reports of off-site shipments. [40 CFR 262.44]

1. Per federal regulation, generators (both large and small) must retain manifest records for three years. Proper record maintenance means that the generator will maintain **on-site** two pages of each manifest that he uses to document off-site shipments. One of these pages — the one that is detached at the time of shipment — will not bear the signature of the owner/operator of the designated facility. The other page will.

2. Similarly, records documenting hazardous waste determination must be maintained on-site for at least three years from the date that the waste was sent to on-site or off-site treatment, storage, or disposal.

3. Retention times of records, per federal regulation, are "extended automatically during the course of any unresolved enforcement action regarding the regulated activity or as requested by the Administrator." [40 CFR 262.40 (d)]

4. Some states may require documents be kept for a longer period of time.

5

Using the Code of Federal Regulations (Titles 40 and 49)

The exhibit included at the end of this chapter provides general summaries of U.S. Department of Transportation (DOT) requirements for shippers of hazardous materials and wastes. DOT regulations are incorporated in both federal and state hazardous waste regulations. These standards are the minimum requirements with which the small quantity generator must comply for the transportation of hazardous waste. The reader is encouraged to review this exhibit and to refer to the referenced sections of Title 49 of the *Code of Federal Regulations* (49 CFR) in order to better understand the structure and wording of transportation regulations.

BASIS FOR THE REGULATIONS

Federal Regulations

As briefly discussed in Chapter 2, all environmental rules and regulations are based upon federal and state statutes. Typically, these statutes direct a governmental agency to administer each statute's provisions. At the federal level of government, the U.S. Environmental Protection Agency (EPA) is charged with administering the Resource Conservation and Recovery Act (RCRA) and a host of other environmental laws. In essence, subject to the intent of Congress, EPA must "make the law work" by developing numerous standards and establishing effective enforcement programs. Thus, while RCRA provides the framework for the development of the regulatory scheme, EPA provides the details and mechanisms that are needed to make it effective.

Notwithstanding the "hammer" provisions of the Hazardous and Solid Waste Amendments of 1984 (a federal law that amends RCRA), the statutory language is typically too general to provide the necessary criteria and guidance that will adequately protect human health and the environment. In a sense, the statute is the skeleton and the rules and regulations promulgated by the EPA are the flesh. Typically, the legislation instructs the EPA on the areas it is to regulate. Hence, EPA is charged with making the legislation workable, and is usually given time frames in which it is to complete certain tasks. The agency's rules and regulations are codified in the *Code of Federal Regulations* (CFR).

State Regulations

Hazardous waste programs are structured similarly at the state level, where the state legislature enacts the law and a state agency administers it. In turn, the administering agency develops regulations which are then published in the state's administrative code.

Small quantity generators (SQGs) who are subject to state-authorized hazardous waste programs must comply with their respective states' administrative code. THIS CAN BE OF CRUCIAL IMPORTANCE TO HAZARDOUS WASTE GENERATORS. It is a serious mistake to investigate and comply with federal regulations alone. Many states publish an equivalent of the *Federal Register* (FR), thereby providing their residents with a single-source reference for changes in regulation at the state level. It is important to realize, however, that states do not immediately adopt newly promulgated federal standards; they adopt the new standards according to highly structured, time-consuming processes. In situations where such a time lag exists, the generator must seek out the CFR and FR for current requirements.

ORGANIZATION OF THE REGULATIONS

As previously mentioned, the myriad regulations developed by federal agencies are referenced in the CFR. This code is readily accessible in book form. CFRs are organized by *title*, such that each administering agency's CFR is given a title that describes its basic purpose. Each CFR title is assigned a specific *number* for ease of reference. In turn, each CFR title is divided into *subchapters* (relating to a broadly defined subject), *parts* (specific areas of a subject), and *subparts*. Naturally, each subpart is composed of numerous smaller parts.

For example, regulations which originate from EPA's rulemaking process are referenced in the CFR entitled "Protection of Environment." The number corresponding to the title is "40." Subchapter I of 40 CFR, which deals with RCRA, contains the hazardous waste regulations in Parts 260 through 280. Part 262 is referenced as "Standards Applicable to Generators of Hazardous Waste" and consists of Subparts A, B, C, D, and E. Each subpart is composed of sections, paragraphs, and subparagraphs.

To summarize, the following list shows the hierarchy by which regulations are organized.

Title/Number
    Subchapters
        Parts
            Subparts
                Sections
                    Paragraphs
                        Subparagraphs

In many instances, subparagraphs are divided into smaller parts. When these divisions occur, attempting to identify a reference in the CFR becomes quite difficult because there is little, if any, "white space" or indentation to provide physical distinction between the various divisions. Consider, for example, a citation like this:

40CFR261.5(f)(3)(v)(B)

A person who is unaccustomed to using a CFR is likely to experience difficulty in finding this reference.

The example above is perhaps best explained in a step-by-step process, as follows:

40 CFR refers to EPA's regulations

261 refers to Part 261 of 40 CFR which is entitled "Identification and Listing of Hazardous Waste" and is found in the volume of 40 CFR which includes Parts 190 through 399

.5 refers to the fifth section which appears in Part 261, Subpart A, entitled "Special Requirements for hazardous waste generated by conditionally exempt small quantity generators"

(f) refers to paragraph f of section 261.5
(3) is a subdivision of paragraph (f)
(v) is a subdivision of paragraph (f)(3)
(B) is a subdivision of paragraph (f)(3)(v)

Hence, any given *section* of the regulations is divided into paragraphs and subparagraphs in this manner:

( ) lowercase letters
( ) whole Arabic numerals
( ) lowercase Roman numerals
( ) uppercase letters

The wording of the regulations is sometimes difficult enough to understand. Remembering the above points will enable one to focus more clearly on the issues at hand.

SUPPLEMENTARY INFORMATION

Edition Dates

In order to maintain some semblance of accuracy in portraying the changes in law and regulatory agencies' rulemaking, CFRs are updated annually. Title 40 is typically revised every July, and Title 49 (Transportation) is revised each October. Edition dates are printed on the book's front cover and cover page, and in the upper right corner of every even-numbered page.

Owning the most recently revised CFR is no assurance of having the most current regulations, especially when dealing with federal policy on hazardous waste management. Changes in the regulations are published daily in the FR, a publication of the U.S. Government Printing Office. The FR provides discussions and codification of rulemaking by many different federal agencies, including the EPA and the DOT. By the time a revised CFR goes to press, it may already be out-of-date.

Fortunately for the regulated community, there are a number of ways to obtain regulatory updates. The EPA, for example, has established the RCRA Superfund Hotline and a hotline for small business. Small quantity generators are encouraged to utilize these hotlines periodically throughout the year. These hotlines are designed not to catch violators, but to provide information that will assist generators in complying. When questioning if an operating practice is legal or not, there should not be a great concern over getting "turned in" by these information sources. However, the SQG is cautioned that since "ignorance is no excuse for breaking the law," he should always attempt to corroborate verbal information by reading a copy of the FR which addresses a particular item of interest.

Authority

The legal authority for the regulations of every *part* of the CFR is referenced immediately *following* each part's miniature table of contents. For example, 40 CFR Part 262 indicates that its legal authority is based on Sections 1006, 2002, 3002, 3003, 3004, 3005, and 3017 of the Solid Waste Disposal Act, as amended by various sections of RCRA. These statutes are codified in Title 42 of various sections of United States Code. Generally speaking, this information is not very practical unless one is attempting to reconcile regulations with legislation upon which they are based.

Federal Register References

Following the text of each *section* in the CFR is a reference to the volume, page number, and date of the FR in which the regulations were initially published.

Notes and Comments

Occasionally, a *paragraph* of a section will be followed by a note which provides comment on the particular standard. A rich example of this sort can be found in 40 CFR 261.33 (d).

Obsolete Provisions

At times, obsolete provisions of a regulation are printed in small type immediately *following* a current regulation. The reader is advised to differentiate carefully between past and present requirements.

Other Information

Other important informational items in the CFR are referenced in the *beginning* of each volume. As described by the U.S. government, these include the following thoughts:

    I. Effective and Expiration Dates. "Each volume of the Code contains amendments published in the Register since the last revision of that volume of the Code. Source citations for the regulations are referred to by volume number and page number of the Federal Register and date of publication. Publication dates and effective dates are usually not the same and care must be exercised by the user in determining the actual effective date. In instances where the effective date is beyond the cut-off date for the Code a note has been inserted to reflect the future effective date. In those instances where a regulation published in the Federal Register states a date certain for expiration, an appropriate note will be inserted following the text."

    II. Obsolete Provisions. "Provisions that become obsolete before the revision date stated on the cover of each volume are not carried. Code users may find the text of provisions in effect on a given date in the past by using the appropriate numerical list of sections affected. . . ."

    III. Incorporated by Reference
        A. "Incorporation by reference was established by statute and allows Federal agencies to meet the requirement to publish regulations in the Federal Register by referring to materials already published elsewhere. For an incorporation to be valid, the Director of the Federal Register must approve it. This material like any other properly issued regulation, has the force of law."

B. "The Director of the Federal Register will approve an incorporation by reference only when the requirements of 1 CFR Part 51 are met. Some of the elements on which approval is based are:
   1. "The incorporation will substantially reduce the volume of material published in the Federal Register."
   2. "The matter incorporated is in fact available to the extent necessary to afford fairness and uniformity in the administrative process. . . ."
C. "What if the material incorporated by reference cannot be found? If you have any problem locating or obtaining a copy of material listed in the Finding Aids . . . as an approved incorporation by reference, please contact the agency that issued the regulation containing that incorporation. . . ."

IV. CFR Indexes and Tabular Guides. "A subject index to the Code of Federal Regulations is contained in a separate volume, revised annually as of January 1, entitled CFR INDEX AND FINDING AIDS. This volume contains the Parallel Table of Statutory Authorities and Agency Rules (Table I), and Acts Requiring Publication in the Federal Register (Table III). A list of CFR Titles, Chapters, and Parts and an alphabetical list of agencies publishing in the CFR are also included in this volume. . . ."

Following is a reproduction of a U.S. DOT form, "Indicators of Hazardous Materials Shipment Violations" (Figure 1).

This is a partial listing of items that you as an *enforcer, shipper, container manufacturer*, or *carrier* may use to spot check for compliance with the DOT Hazardous Materials Regulations. Included in this listing are *indicators only and not necessarily violations* in and of themselves.

The hazardous materials regulations for shippers are found in the Code of Federal Regulations Title 49, Parts 171, 172, 173, 178, 179. These Parts of 49 CFR contain general requirements and communication regulations. This list may be used as a guide when looking for discrepancies or making a compliance inspection. Areas to consider include but are not limited to: *classification, packaging, marking, labeling, placarding, loading, blocking*, and *documentation*. When using this information, remember it is designed to be used as a guide only and does not cover all aspects of the regulations.

I. CLASSIFICATION AND PROPER SHIPPING NAME

A. Improper classification of hazardous materials.

B. Failure to properly classify material having more than one hazard. (Sec. 173.2)

C. Improper description and/or proper shipping name for material being shipped. (Sec. 172.101 and 172.102)

D. Omission of technical name of material following n.o.s. description of material offered for export by vessel. (Sec. 172.203(i))

E. The letters "RQ" not displayed in association with the proper shipping name when required. (Sec. 172.203(c)(2))

II. PACKAGING (CONTAINERS IN GENERAL)

A. Use of DOT specifications containers which are not authorized for the commodity being shipped.

B. Use of containers that are leaking.

C. Manufacturing and marking containers as meeting a DOT specification when they do not meet the specification.

D. Packagings exceeding maximum quantity limitations for materials.

E. Packages improperly marked.

F. Offering for shipment improperly packaged material.

G. Consignee or consignor's name marking omitted from packaging.

H. Omission of identification numbers on packagings. (Sec. 172.301)

III. CONTAINERS (MISCELLANEOUS)

A. STEEL (Sec. 178.80–178.150 plus appropriate section for other areas)

1. Labeled containers (without further overpack) with no DOT specification marking (commonly found are 5 gallon 29 gauge metal pails and 5 gallon rectangular cans).

2. Packages of hazardous materials with temporary repairs.
   a. Damaged, sealed with tape, putty, chewing gum, or screws.
   b. Shipped upside down.

3. Labeled containers in improper condition i.e. dented, rusted or corroded. (NOTE: Some of these are judgemental decisions).

4. Labeled containers on which specification markings are illegible.

5. Labeled reused containers marked "NRC" (look for old date of manufacturer, dents, rust, and paint layers).

6. Labeled reused containers marked "STC" and/or 17C, 17E, and 17H with no reconditioner's marking.

7. Labeled reused containers with a reconditioner's marking that is not a DOT 17C, 17E, or 17H container.

8. Labeled 55 gallon open-head drums with 2 rolling hoops and/or less than 5/8 "ring bolt, non-drop forged ring lugs, and/or "lever lock" ring closures. (Good possibility of non-DOT specification).

9. Imported drums marked as meeting the DOT hazardous materials regulations.

B. FIBERBOARD BOXES (Sec. 178.205–178.219)

1. Boxes with no DOT specification marking when inside packagings are larger than the "limited quantity" exception for the commodity and specification packaging is required.

Figure 1. U.S. DOT form, "Indicators of Hazardous Materials Shipment Violations."

2. Boxes marked with DOT specification markings which are poorly constructed (i.e., gaps, uneven closures, seams and joint separation).
3. If inner flaps do not meet, are fill-in pieces used to fill void?
4. Boxes damaged by water.
5. Improperly closed boxes (*look* for masking tape, cellophane tape, and string).
6. Non-DOT specification *fiberboard* box used in lieu of specification container when required.

C. **POLYETHYLENE CONTAINERS (Sec. 187.21–178.35 a)**
1. Open-head polyethylene containers (used for materials not authorized to be in them).
2. Illegible marked containers.
3. Leaking containers offered for transportation.
4. When poison is shipped, is the container marked POISON?

D. **FIBER DRUMS (Sec. 178.224–178.225)**
1. Non-DOT specification fiber drums.
2. Fiber drums constructed of materials weaker than required by the specification.
3. Use of fiber drums marked DOT-21P without inside polyethylene liner.
4. Using fiber drums marked "STC" more than once for shipping hazardous materials.
5. Fiber drums damaged by forklift truck.
6. Improper markings on containers for the commodity being shipped.

E. **CYLINDERS (Sec. 178.36–178.68)**
1. Re-use of single-use cylinders such as DOT Specification 39.
2. Cylinders in use beyond test date.
3. Cylinders in improper condition:
   a. No valve protection
   b. Bulge in side
   c. Dented or corroded
   d. Defective valve
4. Cylinders re-filled by other than the owner of the cylinder without permission.
5. Cylinders improperly marked (duplication of serial numbers).
6. Cylinders offered for transportation without proper identification of contents.
7. Identification symbols not registered with the Bureau of Explosives or the Department of Transportation.
8. Illegible cylinder markings.

F. **PORTABLE TANKS (Sec. 178.245–178.272)**
1. Name of owners or lessee omitted on tank.
2. No labels and/or placards displayed on tank containing hazardous materials.
3. No identification number displayed on the placard or on an orange panel.

G. **CARGO TANKS (Sec. 178.315–178.343)**
1. Using a cargo tank, marked for one hazardous material, for another hazardous material without proper identification of contents.
2. Improperly marked, e.g. size of marking or not marked in contrasting color.
3. Omission of the marking "QT" (Quenched and Tempered Steel) or "NQT" (other than Quenched and Tempered Steel), when required on cargo tank.
4. Omission of identification number on placard or orange panel.

IV. **MARKING OF CONTAINERS (Sec. 172.300–172.338)**
A. No commodity description (proper shipping name) on the container.
B. No name and address of consignee or consignor on the container.
C. No DOT Exemption number on containers shipped under DOT Exemptions.
D. Container markings not in a contrasting color.
E. Container of liquid hazardous material not marked on outside "THIS END UP" or "THIS SIDE UP."
F. Gross weight not marked on Radioactive Materials packages weighing over 110 pounds.
G. Reconditioned drums improperly marked.
H. USA not included as part of the DOT Specification markings for Radioactive Materials packages destined for export.
I. Portable tanks not marked with proper name of the hazardous material.
J. Omission of identification numbers (when required) on placard or orange panel.

Figure 1.   Continued.

K. Omission of marking of INHALATION HAZARD when required.

V. LABELING (Sec. 172.400–172.450)
A. No labels on the outer container to represent mixed packaging of hazardous materials (materials with more than one hazard – dual labeling).
B. Label on the container not consistent with the hazard class on the shipping papers when appropriate.
C. Use of obsolete labels.
D. Color and/or size of label does not meet the standards of the CFR, Title 49, Sec. 172.407.
E. No label on shipments destined for air transport.
G. Labeling containers not authorized to be labeled.
H. No label on "LIMITED QUANTITIES" offered for air transportation.
I. Less than two Radioactive Materials labels (White I, Yellow II or Yellow III) on containers (two opposite sides).

VI. PLACARDING (Sec. 172.500–172.558)
A. Failure to placard vehicle requiring placarding.
B. Failure to use more than one kind of placard to indicate more than one hazard class of material loaded within vehicle.
C. Freight container containing hazardous material over 640 cubic feet not placarded.
D. Placards not applied to both sides of cargo tank.
E. Placarding material not authorized to be placarded.
F. Omission of identification number (when required) on placard or orange panel.

VII. SHIPPING PAPERS (Sec. 172.200–172.205)
A. No proper shipping name and/or classification of hazardous material entered on shipping papers.
B. Proper shipping name and/or classification abbreviated.
C. No certification for shipment.
D. No wordage for "LIMITED QUANTITY" on shipments excepted from specification packaging and labeling.
E. No DOT Exemption number on shipments moving under DOT Exemption.
F. Color of label indicated in lieu of the proper hazard class.
G. Improper format for hazardous materials description on shipping papers, e.g. HM entries not first, *highlighted* or no *HM column*.
H. No identification number (UN or NA) on shipping paper.

NOTE: This material may be reproduced without special permission from this office. Any comments or recommendations should be sent to:

Training Unit DHM–51
Federal, State, and Private Sector Initiatives Division
Office of Hazardous Materials Transportation
Research and Special Programs Administration
US Department of Transportation
Washington, D.C. 20590

Rev. July 1986

Figure 1.   Continued.

On-Site Storage and Handling of Hazardous Wastes

FACETS OF ON-SITE WASTE MANAGEMENT

How wastes are stored and handled on-site by the small quantity generator has a huge impact on the success or failure of a hazardous waste management program. Proper planning from the beginning is necessary to ensure success, particularly with respect to such factors as economics, safety, reduced liability, and regulatory compliance.

Many factors must be taken into consideration in the development of a waste management program. Wastes must be stored properly, records must be maintained, transporters and disposal methods must be evaluated, and personnel must be properly trained. Of these considerations, the one that involves the most interrelated factors is storage.

STORAGE OF HAZARDOUS WASTE

Following is a list of the different factors involved in the proper storage of hazardous waste. Each of these factors is dependent on the other and must be considered together in the development of the storage plan:

1. Compatibility
2. Packaging
3. Regulatory compliance
4. Segregation
5. Ventilation
6. Climate/environment
7. Space
8. Economics

Compatibility in Storage

In general terms, *compatibility* refers to the ability of two or more materials to exist in close association with each other without the formation of harmful chemical or physical reactions. Incompatibility between two chemical substances can result in one of a number of reactions. Consider the following examples:

$$
\begin{aligned}
\text{acid} + \text{cyanide solution} &= \text{cyanide gas} \\
\text{bleach} + \text{ammonia} &= \text{chlorine gas} \\
\text{water} + \text{lithium aluminum hydride} &= \text{a difficult fire to extinguish} \\
\text{water} + \text{strong acid} &= \text{dangerous evolution of heat and gas} \\
\text{organic material} + \text{strong oxidizer} &= \text{fire} \\
\text{peroxided ether} + \text{shock or friction} &= \text{explosion} \\
\text{isocyanate} + \text{sodium hydroxide} &= \text{violent polymerization} \\
\text{hydrochloric acid} + \text{sodium hydroxide} &= \text{table salt and water}
\end{aligned}
$$

The concept of compatibility, when applied to hazardous waste, refers to:

(1) how chemicals react when in contact with *each other* (see above).
(2) chemicals' compatibility with the *containers* in which they are stored. Incompatibility between a chemical and its container can result in container failure and, ultimately, in environmental damage or personal injury. Some examples include storage of acids in steel drums, storage of certain hydrocarbons in plastic containers, and storage of a pressurized liquid or gas in a weak container.
(3) compatibility with *nearby materials and equipment*. For example, containers of flammable materials should be stored with proper consideration of proximity to heat, electrical connections, and open flames. All flammable containers 5 gallons or larger should be grounded.
(4) compatibility with the *environment* itself. Storing many waste ma-

terials outside may be practical, but storing drums of highly flammable material in dark drums in open sunlight can be extremely dangerous. Weather conditions can have a great impact on container integrity as well. Most shipping containers were not designed to withstand prolonged exposure to moisture, which will corrode steel rapidly.

Economic considerations indicate that materials should be handled in the largest quantities feasible, but materials should only be mixed together when this will not adversely affect disposal cost. Compatibility is not just a concept based on economics, however. Life-threatening situations can be and are caused by failure to take compatibility into consideration.

Safety and regulatory concerns affect the decision process regarding the types of containers used for waste storage. It makes no sense to store acids in a steel drum, because corrosion will surely cause the drum to leak. In fact, steel drums are not permitted for the transportation of most strong acids for this reason. As discussed in Chapter 5, one of the first decisions on the storage of waste is the proper container for shipment. This information is referenced in 49 CFR 172.101. As a general rule, waste materials should always be stored in the containers in which they will be shipped off-site for disposal.

Packaging of Hazardous Wastes

Containers

As previously mentioned, U.S. Department of Transportation (DOT) regulations should be consulted when determining how to properly package waste materials. For most hazardous materials there are more than one permitted container type and size. The quantity of material, requirements of the disposal facility, cost and availability of different containers, and storage space are all factors that must be considered. A reliable transporter or disposal facility can generally provide good advice on this subject. If drums or 5-gallon cans are to be used, it may be possible to store and ship materials in the container the original material was shipped in, which can help to reduce costs. Relatively speaking, disposal costs are much higher for small containers than for larger ones on a per unit volume basis. Handling risks also increase with the number of containers handled, although potential spill quantities are reduced with the use of smaller containers.

Drums. In many cases, reconditioned drums can be used to store wastes. Drums that have been reconditioned and inspected are usually available for approximately half the cost of new drums. Care must be exercised, however, in purchasing such reconditioned drums both from a regulatory and a business perspective.

From a regulatory standpoint, the purchaser of reconditioned drums must realize that drum reconditioners are subject to registration with DOT. For more information on the reuse of drums, consult 49 CFR 173.28 or Appendix X of this book.

From a business perspective, the drum purchaser should be aware of the quality of drum available. Unfortunately, some drum and barrel dealers fail to accurately represent the type or quality of drum they market. It may be advisable to pay a small premium, if necessary, to purchase drums from a reliable vendor. One example of a potential problem is with the purchase of open-head DOT specification 17H drums. These drums should have three "rolling hoops," or raised edges, around the drum. Frequently, drum reconditioners will convert closed head drums into open head drums, and leave only two rolling hoops. These drums may not meet DOT specification, and anyone who ships hazardous waste in such a container may be in violation of the regulations. Despite the potential problems with the use of reconditioned drums, waste containers are rarely returned by the disposal facility, and it does not make sense to purchase new drums if other options are available.

Cans. If storage space is at a premium, containers that can be safely stacked or stored on shelves may be advisable. Examples of such containers include DOT specification 37A (steel 5-gallon cans) or suitable safety cans. Vertical space usage is critical for many small facilities. While storage in relatively small containers represents an efficient use of limited space, it also presents at least two potential problems: increased disposal cost and increased handling risks. It is a relatively simple matter, however, to deal with these potential problems. The key is proper storage.

As previously discussed, it is critically important to address the four factors of compatibility (see p. 72). Especially when utilizing shelving or rack systems, great care must be taken to evaluate the types of wastes that will be stored on any given storage unit. For example, it is unacceptable to store incompatible materials on the same shelf, regardless of their relative positions. Any number of conditions, from container failure to accidental impact, can cause serious damage to life and property.

It is not recommended that wastes be stored with raw materials, since there is always a chance for confusion on the part of either a transporter or an employee. Remember, also, that shelves should always have a lip or border to help keep bottles from sliding off onto the floor. One final note concerning shelf storage: Always bolt any type of freestanding shelves to a stationary object, such as a wall or pillar.

Proper Labeling

Proper labeling is essential. Contamination of raw materials (or waste materials) can be easily avoided by properly labeling all materials and clearly delineating storage areas. Information on handling and potential hazards is easily available, and bulking of compatible wastes is easier at the time of removal. Depending on the agreement with the transporter or disposal facility, the generator may be able to bulk identical or similar waste types into a 55-gallon drum for off-site transportation to a treatment, storage and disposal (TSD) facility. This practice not only reduces disposal costs, but also eliminates the need to replenish a supply of storage containers.

Regulatory Compliance

Though most regulatory agencies at both the federal and state level do not closely regulate the storage of hazardous waste by small generators, accumulation time is always regulated. For generators of between 100 kg and 1000 kg of nonacutely hazardous waste per month, storage is limited to 180 days, or 270 days if the treatment or disposal site for the material is located over 200 miles away. These limits are directed at the storage of wastes in the central storage area of the facility—where wastes ready for disposal are stored.

Where wastes are stored can have an effect on allowable accumulation times; federal regulations allow satellite storage in certain cases. Satellite storage is essentially storage at the point of generation, not in a central storage area. Satellite storage is limited to 55 gallons of nonacute waste in one container and to 1 quart of acutely hazardous waste. A few examples of satellite storage are listed below:

- Spent solvent from tool cleaning operation in a garage
- Spent plating bath from jewelry shop

- Waste ink from print shop stored near the press
- Used solvent from glassware cleaning in a laboratory

Storage of these quantities is allowed for one year by the Resource Conservation and Recovery Act (RCRA), though state regulations may vary. Once the waste has been removed to a central waste storage area, an accumulation date should be recorded on the container, and normal rules for shipment off-site are in force. Small quantity generator status is maintained until the 1000-kg (or 1-kg acute) level is reached.

Segregation of Wastes in Storage

Compatibility is the prime factor in determining segregation of wastes in storage. Segregation of wastes figures largely in the decision as to off-site disposal method. Economics and environmental concerns both suggest that treatment and incineration are the two most sensible disposal methods. How wastes are mixed on-site, however, can dictate the cost and suitability of wastes for particular disposal options. The most practical demonstration of this is the disposal of waste solvents. Following are fairly standard criteria for the fuel recovery of flammable solvents, which represents the most cost-effective (read *cheapest*) disposal method for a waste material:

high BTU value (> 20,000 BTUs)
< 2% total chlorine
< 2% total sulfides
low odor
< 5% water
< 5% total solids
no PCBs, pesticides, heavy metals, or highly toxic materials

As you can see, the mixing of chlorinated and nonchlorinated solvents will render a waste unsuitable for fuel recovery. To get a high chlorine content material to burn, clean fuel must be added, and gas scrubbers must be utilized to collect and neutralize hydrochloric acid, a product of chlorine and the combustion process. Chemical compounds such as mercaptans and sulfides can cause a problem for disposal facilities because of the unpleasant odor associated with these materials. Most neighbors, industrial or residential, can become considerably less friendly when exposed to the odors from these compounds.

The presence of a high solid content can present two different difficulties for a fuel recovery program. First, the waste can present a material handling problem by being difficult to pump. Sludges and semisolid materials are much harder to handle with conventional equipment, and specialized equipment for handling these materials is more expensive. Many fuel recovery programs utilize cement kilns which were not designed for high solid content.

A second problem is the additional ash generated from the burning of these materials. This ash must be transported to a landfill for disposal, which increases the cost of the overall operation considerably.

From the standpoint of chemical treatment or neutralization of a waste material, it is usually desirable to separate organics from inorganics. Common acid and alkaline waste streams can easily be treated at many facilities by diluting, mixing the high and low pH materials together in predetermined concentrations, and then dewatering the resulting mixture. Filter cake which may in many cases be nonhazardous can then be cheaply and safely landfilled. Organic materials, however, can affect the process adversely and may make a material unsuitable for treatment at many facilities.

Another concern in the mixing of wastes on-site is the safety factor. Unless the materials are tested ahead of time for compatibility, the possibility of a dangerous reaction or explosion is present. Accidental or deliberate contamination of one waste stream may entirely change its characteristics when mixed with another stream.

Additional information on disposal options will be covered in a later chapter, but it is imperative for economic and safety reasons that careful consideration be given to any mixing of waste streams.

Ventilation

Any area used for storage of chemical wastes or any other hazardous material should be well ventilated. Highly volatile organics in particular can present a serious health hazard in storage. Exposure limits for many raw materials are regulated under the federal Occupational Safety and Health Administration (OSHA), but little has been done to date to regulate exposure to hazardous wastes in the workplace for small quantity generators.

For many small quantity generators, the answer to the problem of ventilation is to store materials outside in sheds, which provide free air movement. Regardless of the apparent adequacy of the ventilation system, it is vitally important to provide proper respiratory protection for employees who may be exposed to vapors or fumes. A half-face air purifying respirator which is outfitted with organic vapor/acid gas cartridges generally provides adequate protection in work areas where common organic solvents or mineral acids (eg., sulfuric, chromic, nitric, or hydrochloric) are used. Respiratory protection must always be provided per the standards issued by OSHA or an equivalent state agency. Federal standards for respiratory protection are referenced in 29 CFR 1910.134.

Adequate ventilation of waste storage areas is also important because it even provides a degree of protection to individuals who are not directly involved in waste handling. In the event of container failure, an adequate ventilation system will effectively route the vapors out of the building, thereby reducing potential exposure to other employees. The key in establishing a ventilation system — no matter how simple — is to ensure that the air pressure surrounding the waste storage area is greater than the air pressure in the storage area. In a sense, such a condition will prevent potentially contaminated air (from the storage area) from mixing with clean air.

The importance of ventilation is graphically highlighted by the result of improper storage of flammable solvents. A source of ignition (flame, heat, or sparks) in an area where vapors from flammable solvents compose only 2% of the air can cause fire or explosion. Consider the amount of vapor that accumulates in an unventilated storage closet or cabinet on a hot summer weekend. Consider too, that vapors from flammable solvents are heavier than air and that the vapors from a broken 1-quart container can travel as much as 100 feet in several seconds. It is not very difficult to imagine the destruction that can occur if someone causes a spark, perhaps from dropping a steel object on a concrete floor, or if someone initiates a static charge to arc through the air as he walks on a nylon carpet.

Unquestionably, the installation of an intrinsically safe exhaust fan or other type of ventilation can reduce the risk of fire and explosion. However, regardless of the type of ventilation provided, care must always be exercised to prevent exhausted air from reentering the building through doors, windows, and air intakes on the building's ventilation system. When properly installed, flammable storage cabinets and other specially designed systems provide good protection against vapor accumulation.

Climate/Environment

As storage space is a potential problem for nearly all small generators, weather conditions can frequently be an important factor in determining storage conditions. Heat, cold, moisture, and wind can adversely affect the safe storage of all chemicals. A steel drum left outside in a northern winter or a wet spring will rapidly deteriorate, rendering it unsuitable for storage or shipping. Similarly, weather conditions can cause deterioration of labels, thus increasing the risk of improper handling and disposal. This deterioration of labels can take place in a matter of days or weeks.

The ability to properly dispose of a waste material can be compromised in freezing weather when the material is a liquid with a high water content. Under normal conditions, an aqueous solution might be suitable for treatment as a waste water. As a solid, however, it cannot be pumped, and thus can present severe problems in the winter. More than one generator has had to wait for the spring thaw to dispose of liquid waste materials stored outside.

Flammable materials stored outside in drums during a hot summer present a different problem. A steel drum painted a dark color can easily rise to temperatures above 100°F. Pressure buildup from high temperatures can damage a container's integrity if venting is not provided. A pressure buildup can also result in a "spraying" of hazardous waste to an individual who unwittingly opens a drum to add more waste.

Many materials go through a freeze/thaw cycle in changing weather conditions. This too causes metal stress, leading to leaking containers. Also, adverse weather conditions will attack labels and other types of identification attached to a container. If a number of different materials are stored in adjacent containers, it might prove difficult to easily identify the wastes. Costs for the identification of unknown materials can cost hundreds of dollars.

Other weather-related problems, including contamination of a waste with rainwater, contaminated soil disposal as a result of leaks, or degradation such as polymerization of a material, may contribute to higher costs for disposal.

If waste materials must be stored outside, they should always be covered by a roof or tarpaulin, away from direct sunlight. Drums or other containers should be stored on pallets, or kept off the ground in some other way. Secondary containment should be provided in the event of a leak or spill. Ideally, the containment device (berm, dike, tray, liner, etc.) should have the capability of holding at least 10 to 15% of the total volume of the stored containers. While there are few actual federal or state standards for the storage of hazardous waste by small quantity generators, the costs and safety factors associated with improper storage should be incentive enough to store materials properly. This is also obviously true of storage inside a building.

Space

For any small businesses, the problem of available space has to be among the most significant issues addressed when considering any change in procedures. With the cost of space representing a large part of any small business budget, any new space requirements need to be carefully reviewed. The on-site storage of hazardous waste is a particular problem, in that all of the previous factors affect the type of space required and the amount that must be stored at any one given time.

Suggestions for conservation of space include the use of shelving, the bulking of materials into larger containers, more frequent removal of waste, waste minimization (in-house reclamation or recycling), and use of stackable containers.

Economics

Economic considerations for the small quantity generator of hazardous wastes involve more than careful planning of expenditures. All of the previous concerns must be addressed; many ways to save money without substantially increasing liability are included.

Choosing a Waste Management Company

Careful selection of disposal methods, waste management brokers, and transporters can certainly reduce costs. Maintaining control over these decisions is important, since, in some cases, a generator may actually be dealing through a succession of brokers without actually being aware of it. Few transporters or brokers own or operate their own disposal sites, and they may be going through other transporters or brokers to gain access to sites. Some disposal sites limit the number or type of companies they will accept waste through, either by controlling transportation, limiting schedules, or dealing only through a specific marketing representative.

It is important to know exactly what steps are involved in getting your waste materials into a disposal site, and how many different companies are involved. While few small generators are in a position to deal directly with a disposal site, minimizing the number of middlemen will certainly reduce the costs.

Transportation rates can also vary considerably, depending on the type and condition of vehicles, distance to disposal facilities, size of the fleet, number of terminals, and labor costs. Frequently, a small company with only a few trucks can offer a lower rate but less timely service. The greater the distance to the disposal site, the more likely that a larger transporter (one who is permitted in a greater number of states) will be required. The number of trucks and terminals can greatly affect rates, because backhauls are more often available to larger companies.

Additional information on choosing a waste management company will be provided in Chapter 7, but this selection is greatly affected by on-site management efforts. The limiting of liability and reduction of costs begin with a good system of on-site handling and storage of hazardous wastes.

Effect on Management Methods

Methods chosen by the small quantity generator to effect on-site storage and handling of hazardous wastes are laden with economic issues. This is an unfortunate yet necessary evil, given the importance of cost controls for small businesses. While safety is frequently a much more visible consideration for large generators, small generators must in many cases compromise if they are to fully comply with regulations. To put the costs of proper storage in perspective, consider some specifics of space availability, material requirements, and personnel commitments that are required to properly and safely manage wastes on-site.

Storage Space.   Small generators that have little or no storage space available for waste materials are forced to "make do." A number of potential results of severe space limitations include:

- disposing of less than economical quantities of waste at one time
- combining waste storage with raw material or equipment storage
- keeping waste "out back," where climate factors may contribute to rapid degradation of containers, and thus to potential contamination of the environment
- illegal dumping

Storage space is thus a key factor in operating a good waste management program. Expansion of the physical plant may not be desirable from a short-term economic viewpoint, but long-term liability must be considered.

Materials and Supplies.   Material and supply requirements for a waste chemical storage area are contingent on many factors. Laboratory waste will be considered in Chapter 9. Waste from other small generators is typically stored in drums or cans meeting DOT specifications. From a material handling standpoint, and considering the relatively small quantities involved, drums are much more desirable than tanks. For quantities less than 1000 gallons, tank storage is uneconomical. Storage of greater quantities will force the generator to exceed waste accumulation times specified by regulation for small quantity generators. On the other hand, drum storage of hazardous waste offers the small quantity generator several advantages over tank storage, including:

- Ready availability of drums through local suppliers, at low cost
- Greater flexibility in terms of reorganizing or relocating operations
- More flexibility in managing different waste streams
- Reduced risk in the event of container failure
- Greater cost control associated with short-term storage

Drums are readily available in a variety of sizes, and few supplies other than the containers themselves are required. Tools for opening and closing the drums (bung wrench or socket wrench for most drums) should be kept on hand, and all containers of flammable liquids should be electrically grounded.

Personnel Requirements.   All waste management programs need a responsible person to serve as hazardous waste manager. The fact that this may be a part-time position in many businesses should not diminish the importance of exercising care in the choice of a manager. Regardless of the size of a business or organization, individual accountability is key to the success of a waste management program. The higher the level of responsibility this person has in the organization, the greater the chance for a successful program. Simply put, support from senior management is mandatory. Too often in small businesses the responsibility for handling waste falls to someone without the power to develop and enforce internal policies. Without these policies, chances for problems in a waste management program are greatly increased. Chapter 8 provides more insight into personnel selection and accountability.

The hazardous waste manager should be trained in hazardous waste management, regulatory compliance, and safety and should have some experience in managing personnel. All of this training is available commercially, and more and more colleges are beginning to offer classes in environmental science and related fields.

For most small generators, once a program is established, it is not necessary to commit a great number of manhours to waste management. In many cases, the hazardous waste manager may serve in this capacity on a part-time basis. It is far more important to take the initial steps to properly train someone in the management of the program, or to hire someone with experience. If the decision is made to subcontract all disposal services, it is still necessary to evaluate and monitor transporters and disposal facilities. With the emphasis that is currently placed on compliance and liability by enforcement agencies, the media, and the public in general, failure to show responsibility for the management of an organization's waste can have serious implications. Especially in a business that is not incorporated, such as a sole proprietorship or a partnership, the consequences of poor waste management practices readily translate into *personal liability* for the business's owners and executives. An honest effort in managing hazardous waste is far superior to no effort at all. Liability reduction begins with establishing *personal accountability.*

7

Limiting Liabilities

INTERNAL ENVIRONMENTAL AUDITS

Purpose

The most obvious purpose in completing an internal environmental audit is to ensure that the facility is in compliance with regulations. An audit consists primarily of an inspection of the facility and a review of the inspection in light of regulatory requirements. Necessary corrections should be implemented immediately. Ideally, such an inspection will also increase safety awareness and expose employees to more information on the regulations than they might ordinarily receive.

An environmental audit must, above all, be objective. It may be difficult for employees of the facility to maintain this objectivity since familiarity with the location may cause something to be missed that might be more obvious to an outside observer. On the other hand, a full understanding of the logistics of the facility can be a plus. Understanding procedures, financial limitations, employee awareness, and other factors may give insight into how minor changes can affect compliance. It might be advisable to perform an audit by committee, with at least one outside observer participating to ensure objectivity.

For a small business lacking adequate personnel and expertise, a trade association may be a source of both information and advisors to participate in such an audit by committee.

The audit inspection requires some knowledge of regulatory requirements as they apply to the specific location. Figure 1 represents an audit questionnaire that addresses nearly every possible facet of hazardous waste compliance. Some questions may not be applicable to small generators, but these are included because of the wide variety of different state regulations.

Audit Follow-up

An environmental audit points out problem areas at the facility and is only useful if these problems are corrected. Careful consideration should be given to different ways to solve these problems. It may be necessary, for instance, to seek outside training sources if the expertise is not available in-house. There may be more than one way to rectify a particular deficiency, and with the normal economic concerns of small businesses it is important to consider all of the possibilities. For example, it may be possible to arrange for consulting through a local chamber of commerce or trade association. Through either of these organizations, members can submit their individual questions to their local chapter. In turn, the chapter leader would forward the questions to a guest speaker or consultant. A general member meeting can then be arranged where the speaker can lecture and provide time for a question and answer session. Local colleges and universities are also a possible source of information or consulting services.

CHOOSING HAZARDOUS WASTE MANAGEMENT COMPANIES

The choice of hazardous waste management companies and facilities is one of the most important decisions facing a small quantity generator. With the general shortage of available staff for environmental auditing and site investigation, small generators need to depend heavily on outside management services. The selection of these services should reflect any concerns over long-term liability anticipated by the generator.

Investigating waste management and disposal firms involves using

Facility name _____
EPA number _____
Date of Inspection _____ Time Start _____ Time finish _____
Inspector(s) _____
Location _____
County and municipality _____

1. Types of waste generated by Hazardous Waste number _____
_____

2. Average total quantity of waste generated per month _____
   per year _____

3. Waste handling methods (check all applicable categories)

   On-site treatment ____ storage ____ disposal ____ recycle ____ reuse ____ reclaim ____
   Off-site treatment ____ storage ____ disposal ____ recycle ____ reuse ____ reclaim ____

4. Who transports waste off-site? _____

5. Are transporters fully permitted in all applicable states? _____

6. Has authorization been received for waste shipments to a TSDF? _____

7. Is a manifest file maintained? _____
   Who fills out manifests, generator or transporter? _____
   Have the appropriate state manifests been used? _____

8. List all previous manifest discrepancies
   _____
   _____

9. Have all discrepancies been corrected? _____

10. Are all waste containers on site properly marked? _____
    Are accumulated dates listed? _____
    Are all containers in good condition? _____

11. Are containers offered for shipment properly labeled according to U.S. DOT specifications? _____

12. How long are wastes stored on-site? _____
    What is the transporter's lead time for shipments? _____

13. Are proper records kept? _____ How long are they kept for? _____

14. How often are state or federal reports sent? _____

15. Are exception reports properly filed? _____

16. Are there any spill reports filed? _____
    Were conditions corrected satisfactorily? _____
    Is a spill response plan available? _____

17. Is a Preparedness, Prevention, and Contingency Plan required? _____
    If so, has it been implemented? _____

18. Is a waste minimization plan required? _____
    Has one been developed? _____

19. Has the required training of employees who handle hazardous waste been performed?
    _____
    Under what circumstances is additional training performed? _____

Figure 1.   Hazardous waste generator inspection report.

all relatively accessible resources. Following is a list of the different factors involved in investigating transporters, waste brokers, and disposal facilities, and methods of evaluation for each factor.

1. Transporters — permits, vehicles, references, experience, personnel, insurance, financial stability, lead time, cost
2. Waste brokers — experience, references, personnel, cost
3. Disposal facilities — location, materials handled, permits, notices of violation, financial stability, personnel, insurance, approval process, cost

Transporters

Transportation of hazardous waste is the most visible factor in the overall scheme of hazardous waste management. Vacuum trucks, vans, tractor-trailers, flatbed trucks, and stakeback trucks are all utilized for waste transportation, and any of these vehicles may be encountered on the nation's highways at any time. The permit requirements for *waste* are quite different from those for hazardous *materials*, however.

Permits

Transportation of hazardous waste is regulated not only by the U.S. Department of Transportation (DOT) and U.S. Environmental Protection Agency (EPA), but also by state and local agencies. Most states have permit requirements for transporters, and these range anywhere from a $25 permit fee to a $60,000 cash bond. Some states have licensing requirements for drivers, and an effort is taking place to require the licensing of all drivers who transport hazardous materials nationwide. Other factors that may be involved in permitting include registration of specific vehicles, listing of all disposal facilities in a state on the permit, and inspection of vehicles. As a result of the high cost and difficulty in permitting in some states, few transporters, even large ones, have permits for more than 10 or 15 states. The location of disposal facilities and customer base usually determine the states for which a transporter will obtain permits.

Many transporters carry regulated hazardous materials in addition to waste. It is important to note that the permit requirements for waste are substantially more detailed. Although the hazards associated with hazardous materials (nonwastes) are frequently greater, there is more perceived risk involved with waste. The only logical reasons for this are the possibility of illegal dumping of waste by a transporter and the "evil" reputation of hazardous wastes associated with environmental problems.

Permits can be checked by calling or writing to state environmental protection agencies. A list of these agencies with their phone numbers is shown in Appendix I.

Vehicles

Vehicles used to transport wastes come in every possible shape and size. Evaluation of a transporter's capabilities includes determination of whether the proper types of vehicles are available for your waste materials. Generators of drummed waste, for instance, would not ordinarily consider using a transporter that specializes in vacuum trucks. Generators of laboratory waste would not usually have their labpacks transported by a firm that has only tractor-trailers.

The condition of a transporter's vehicles is a good indication of whether the company should be utilized. Companies with extremely old, decrepit trucks may be inexpensive, but frequent breakdowns or other problems may create additional costs and liability to the generator. Violations of state or federal regulations by a transporter may result in difficulties for the generator.

References

Perhaps the best method of selecting a suitable transporter is to seek the recommendations of other generators in the approximate vicinity of your facility. A long list of happy, satisfied clients is the best endorsement for a transporter. While state agencies can verify permits and report violations to a potential customer, it is the firm's clients who can best detail such factors as reliability, experience, and professionalism. Most transporters will be happy to supply references upon request, but investigation of manifest files at a state environmental agency is the most effective way to obtain a customer list that will not necessarily consist of a transporter's favored clients.

Experience

As the old adage goes, there is no substitute for experience. Transportation of hazardous waste involves so many different regulatory requirements and additional factors that experience must be considered. Since most transporters also act as brokers, the selection of a disposal option and facility is frequently a part of the transporter's services. Giving the responsibility of making disposal decisions to another firm has risks that can only be minimized by careful evaluation of the firm.

Checking references is the first step in evaluating experience, but knowing what types of projects the firm has completed and how long they have been in business can be important in establishing reliability as well. Involvement with large emergency response projects, for example, would be a good indication that the firm can handle its own emergencies should they arise. Transporters that have handled large cleanups with varying types of waste would indicate a varied background and proficiency with a number of different waste types. It is best to ask a transporter for a list of projects they have completed in addition to references on routine work for clients.

Personnel

Professional appearance of personnel is a good indication of the proficiency of a company. Care for personal safety and hygiene by a driver translates to care about a job well done. The attitude of a driver on any given day is important to the customer who must put faith in the driver to handle his waste materials properly.

Knowledge of regulations and procedures by a driver or other company personnel should be documented by annual training and testing. Evaluation of personal driving records is often overlooked by transporters; it is well documented that over-the-road drivers frequently carry licenses in more than one state— illegally—to hide poor driving records.

It is important for a transporter's sales and administrative personnel to present a professional appearance and attitude. The first impression that a salesman is knowledgeable and cares about the customer's liability cannot be underestimated in the decision-making process. If the transporter is acting as a broker and dealing with disposal facilities on behalf of the generator, he should be familiar with approval procedures and policies. Even knowing the proper way to take a waste sample and having the proper supplies and equipment can be an indication of the reliability of the firm.

While it is the generator's responsibility to label wastes properly, complete the waste manifest, and offer placards for the vehicle, all these services are offered frequently by transporters, providing more hints about the experience level of personnel. It should always be determined ahead of time who is to execute these responsibilities and to verify that they are being done properly.

Insurance

While the availability and cost of insurance varies considerably, transporters must have insurance to satisfy legal requirements and to help reduce the liability of the generator. One million dollars of general liability insurance is a common requirement for permits, although $500,000 may be sufficient in many cases. In certain cases, such as large tank trucks or when especially hazardous materials are to be transported, $5 million coverage may be required. A company's ability to secure required insurance is indicative of its financial stability. A few large transporters may be able to get larger amounts, but these companies are usually self-insured. At any rate, generators should always verify that the transporter has up-to-date policies for general liability and workers' compensation. A certificate of insurance is usually supplied upon request, and it may be possible to be named as an additional insured for additional cost. Your own insurance broker can give you advice on this procedure.

Financial Stability

A transporter with severe financial problems can represent a risk to a potential customer. Obvious difficulties may show up in the condition of vehicles, the attitude of employees, and the status of insurance and permits. While it may be difficult to evaluate financial stability in a transporter, an effort should be made to ensure that the company has the financial resources to properly handle waste materials and to deal successfully with problems such as spills and other emergencies. A credit check on the company may help in evaluating financial resources.

Lead Time

Few small generators consider lead time when they are evaluating a potential transporter. For those facilities that need to move waste materials off-site promptly to maintain regulatory compliance, long lead times can present substantial difficulties.

With a 90-day time limit for the storage of waste in many states and limits of either 180 or 270 days under federal statutes, it is imperative to schedule removal in a timely fashion. A transporter who offers low rates but cannot deliver on time is not worth the perceived savings. Lead times in the hazardous waste industry range between two days and two months, depending on the type of work involved. A straight pickup of bulk stream wastes by vacuum truck should be available within a week. Pickup of drums can normally be accomplished within two weeks.

Cost

While cost is frequently the most obvious factor in selection of a transporter, it should not be the only factor. Costs can vary considerably based on the type of vehicle required, the transporter's volume of business, profit and overhead margins, lead time, and perceived level of competition. It is important to carefully evaluate the way a job is quoted, since transporters may charge by mileage or travel time and may add equipment usage, demurrage, labor, and other costs.

It is best to get a firm quotation, if possible, for any work. If an estimate is offered, be sure that you are aware of any factors that may add cost. If there are multiple bidders for work, be sure that the additional evaluation factors are considered, and try to get a not-to-exceed rate whenever possible. This may result in a higher estimate for work, but will prevent "surprises" that can substantially increase costs.

There are a number of ways to make the transporter's job easier and reduce costs. These include providing easy access to the waste containers, labeling containers in advance, and helping to load the vehicle.

Waste Brokers

An experienced broker can save a generator a great deal of time and money, but the service provided should be carefully evaluated. A broker is a consultant who offers some or all of the following services:

- Evaluation of the waste generation process
- Identification of the waste material
- Regulatory classification of the waste
- Selection of the appropriate disposal or recycling methods
- Arrangements for approvals at authorized facilities

- Transportation by contract or subcontract
- Generation and administration of required paperwork

Frequently brokers specialize in one or more of these areas and use a substantial network of resources to complete the services required by the customer. It should be noted that the broker is making all of these decisions on behalf of the client and may not share the same liabilities as a generator, transporter, or TSD facility. It is important for the customer to become as knowledgeable as possible regarding state and federal regulations in order to recognize discrepancies in the performance of a broker. A full service broker should be able to quickly identify serious compliance problems and be able to keep his customers abreast of important regulatory changes. It is not advisable to use a broker who does not provide complete answers regarding handling of waste materials.

The advantages of using a qualified broker include technically competent management of waste handling and disposal, prompt response to customer needs and reduction in the necessary interaction with regulatory agencies and the hazardous waste industry. A broker can sell his services as a consultant, charge a management fee, or add a commission to direct billings. The customer should regularly assess a broker's performance and make inquiries into competitive pricing. The fast-paced nature of the hazardous waste disposal industry tends to promote large variances in pricing for identical services.

Experience

Experience and credibility are the most important things a broker has to offer. It is important to investigate the broker's background and reputation. Evaluate at least three credible references; these references should closely reflect the desired services. Ask for additional references if the nature of work performed for those contacts is not sufficient for proper evaluation. For example, a consultant who handles extensive permitting and design services for a waste treatment system may have no experience with arranging disposal of miscellaneous lab chemicals.

There are also many brokers possessing excellent selling skills but little technical ability. Make sure your assessment of a broker goes beyond a personality evaluation of the salesman. A good waste broker has experience with a wide variety of different projects, from the disposal of process wastes, to laboratory wastes, to environmental cleanups. Evaluating the types of services performed and projects completed can assist the generator in making appropriate selection of a broker.

The importance of selecting a qualified broker is directly related to the liabilities of being a small quantity hazardous waste generator. A good broker is familiar with the myriad of regulations and can help you take all of these factors into consideration. Your broker does not share these liabilities with you, but can provide the necessary support to keep you advised of potential problems.

Personnel

Technical training is most important for a waste broker, particularly when waste must be characterized prior to disposal method selection and facility approval. Proof of training for personnel should be required, especially since most states require no permits or registration for waste brokers. Personnel should be familiar with hazardous waste transportation and disposal regulations and should be carefully evaluated with these considerations in mind.

Cost

Brokers believe in charging market price for services offered. It is to the advantage of the customer to clearly define the required tasks or services to obtain the best value. On the average, brokers mark up rates from 20% to 30% above their direct costs. Depending on the nature of the service, this fee can be cost-effective for many SQGs.

As previously mentioned, brokers can charge either a straight fee for services or mark up subcontractor rates. Costs for identical services can vary considerably, depending largely on the overhead and profit margins. Evaluating all of the previous factors is important to determine whether costs are reasonable.

Of obvious importance is the determination of exactly what services are being offered. Generators may be advised to handle dealings with transporters and disposal facilities directly, but if the experience is not available in-house to make informed decisions, the use of a qualified broker is a good investment. Of particular note is the possibility that costs can be reduced if the broker is able to arrange similar services with several consecutive customers. "Milk runs," for example, combine your waste materials on the same truck with other small generator material and allow for discounts in transportation prices.

Disposal Facilities

Once the best disposal method has been determined (see Chapter 10), the selection of a facility that is permitted for the particular method must be made. While certain technologies are not available in many regions and are appropriate for only certain materials, there are still a number of factors that must be considered to determine the disposal facility to be utilized.

As we have seen, the selection of a disposal facility is frequently made by either a broker or transporter. If the generator has the personnel available to evaluate disposal facilities and deal with them directly, it may result in a cost savings. Some facilities offer quantity discounts to brokers and transporters, and others may have special working relationships that make it advisable to deal through an intermediary. Nonetheless, selection of disposal facilities is the ultimate responsibility of the generator, and how waste is handled for ultimate disposal has a significant impact on long-term liability for the generator.

Location

The geographical location of a potential disposal facility becomes important when transportation costs are considered. A generator in Florida may not want to use a disposal facility in California and vice versa. Transportation of a truckload of material from coast to coast may range as high as $10,000 to $12,000. The cost for the same load for a local delivery will be only several hundred dollars.

Materials Handled

Most disposal facilities are suitable only for a limited number of different waste types. Once the most appropriate disposal method has been determined, permits of various facilities can be checked to determine acceptability. A good waste broker or transporter will also know what types of a waste a facility can accept. Many state Resource Conservation and Recovery Act (RCRA) programs require generators of hazardous waste to receive written authorization from the intended disposal facility indicating that the disposer has both the capacity and ability to accept each waste stream being offered by the generator.

Short of delving into a full-scale review of a facility's operating records (a difficult task at best, depending on the size of the facility), there is little a generator can do to verify the disposer's capacity to accept a particular waste stream. In essence, he must rely on the integrity of the disposal facility's management. However, verifying the types of waste that a disposal facility can handle is a completely different matter. In this area, the generator has several tools at his disposal, including:

- Freedom of information (FOI) at the EPA regional level. A typical FOI request would reference basic information about the facility (name, location, etc.) and would specifically request the EPA waste types that the facility can accept.
- Part A application of the facility, which specifically indicates the EPA waste types that the facility can accept.
- Review of the facility's permitting records at the central or regional offices of the authorized governmental agency of the RCRA program.

Permits

Disposal facilities may be permitted by the U.S. EPA, by the state, or both. States with RCRA authority to regulate waste typically issue their own permits. At this time, most disposal facilities are still operating under interim status. The issuance of final permits cannot be made in most states until an RCRA Part B permit application has been submitted, reviewed, and revised; financial requirements have been met; and public hearings have been held.

There is a substantial backlog of applications to process in many states, and final permitting of many facilities may not be complete for years. State environmental agencies and EPA regional offices can advise generators of the status of permitting of any facility.

Notices of Violation

One way to determine the acceptability of a disposal facility is to investigate any notices of violation that have been filed with the regulatory agency responsible for administering the rules that govern the facility. Facilities should reveal these violations, and a determination should be made about the seriousness of violations. Due to the incredible number of regulations facing the industry, some paperwork violations are almost unavoidable. While an excessive number of paperwork violations can indicate a problem, more serious violations should immediately cause a red flag to go up in the facility evaluation process. Examples of serious violations and other problems include improper disposal, acceptance of wastes for which the facility is not permitted, and poor security of a facility. Administrative orders, consent decrees, and settlement letters are other enforcement mechanisms used by regulatory authorities to compel facilities to comply with regulations. All such documents can be found in the regulatory agency's compliance section.

Most facilities have limits on the number of containers they can store prior to disposal, and visual observation of several thousand containers is usually not a good indication of a properly run facility.

Financial Stability

The costs of operating and maintaining a waste disposal facility can be enormous. Regulatory compliance alone contributes substantial costs to running a facility. It is imperative that the facility have the required financial mechanisms for closure and postclosure care in place. This is most important when the facility is involved with land disposal, which requires perpetual care.

The condition of equipment for treatment or disposal of wastes is an important evaluation factor. An incinerator with frequent downtime, for instance, cannot run efficiently. Wastes being stored prior to disposal can pile up quickly, and any unprocessed waste leaves the generator with continued liability for its proper disposal should the facility fail.

Generators should make sure that the facility has enough financial reserves to handle unforeseen difficulties such as equipment breakdown, environmental accidents, and severe weather problems.

Federal law requires that treatment, storage and disposal facilities (TSDFs) carry environmental impairment liability insurance for both sudden and nonsudden releases. Typically, these regulations require a minimum coverage of $1 million per occurrence and a $2 million annual aggregate for sudden (accidental) releases. For nonsudden releases, coverage in the amount of $3 million per occurrence and a $6 million aggregate is required, per 40 CFR Part 264 and Part 265 Subpart H. TSDFs can supply a copy of their applicable insurance certificates.

Personnel

The professionalism of disposal facility personnel can tell the generator a great deal about the facility itself. Of particular concern should be the employees with responsibility for regulatory compliance and those responsible for technical approvals of waste. With regulations changing day to day at many different levels, maintaining compliance is a full-time job.

TSDFs are scrutinized constantly by inspectors, and even paperwork violations can result in fines and the associated adverse publicity. Some states will place inspectors at a TSDF on a nearly full-time basis to ensure compliance, particularly if there is a history of problems at the site. Having a compliance specialist on staff to handle regulatory updates, perform training of employees, and work with inspectors is necessary at most facilities. At smaller facilities, it is generally the facility manager who has regulatory compliance responsibilities. Determining the level of experience of the individual or individuals responsible for compliance can provide a great deal of information on the quality of service offered by the site.

Technical approvals of waste at a TSDF involve analytical testing as well as a thorough knowledge of the facility's capabilities in treating or disposing of waste. No evaluation of a facility is complete without assuring that the facility technical manager (or site chemist, in some cases) has a high level of experience with both analytical procedures and the equipment necessary to properly evaluate a waste stream. The laboratory involved in waste stream evaluations should have the proper equipment to handle these responsibilities in a timely fashion without sacrificing accuracy. An improper evaluation can jeopardize the facility's safe operation (and has done so in some cases).

If possible, waste should be thoroughly evaluated by the generator prior to sending a sample to the disposal facility for approval. Knowledge of the facility's allowable parameters for such factors as halogen level and presence of heavy metals can help the generator determine in advance if the disposal method is suitable.

The professional appearance of other facility employees can tell a great deal about how a facility is run. The proper use of safety gear and the housekeeping appearance of the facility are additional obvious factors that should be taken into consideration during a site evaluation.

Insurance

Evaluating whether the facility has the appropriate insurance is necessary for the generator. When looking at a copy of the insurance certificate, be certain that the various policies are currently in force and that the limits meet regulatory requirements. State agencies can provide information on the appropriate amounts required.

Approval Process

All disposal facilities are required to have a process by which new waste streams are evaluated. A lengthy approval process is not always necessary, but should reflect the allowable limits of the facility's operating permit. If neither samples nor analytical data are required (except for some facilities handling laboratory waste), there is a much greater chance of wastes being accepted that are not appropriate. While getting waste streams approved is frequently a time-consuming proposition, it is necessary for the generator's protection as well as the facility's. Liability for an accident involving materials not as described can belong to the generator as well as the facility.

There is also the possibility of greatly increased cost for the handling of off-specification waste, or that the shipment may be rejected. This will increase the cost for transportation.

Costs

All of the previous factors will necessarily be reflected in the costs for disposal at a facility. While these costs may be the most visible factor in facility evaluation, it is a serious mistake to consider only cost. The liability of the generator must be considered on both the short term and the long term. Rates for disposal can vary widely as a result of operating costs, geography, market considerations, volume, and profitability of a facility.

8

Enforcement

Small quantity generators (SQGs) are not normally high on the priority list for state or federal inspections. It is not unusual for a SQG to never undergo an inspection, though most states claim it is their objective to eventually inspect all generators.

There are certain instances, however, where an inspection is more likely. If a spill or any other environmental incident is reported by the facility, the chances for an inspection are greatly increased. A complaint by a present or past employee concerning unsafe or illegal practices involving the handling of hazardous waste will surely increase the likelihood of an inspection. Such a complaint may also give the inspector impetus for a more complete review of documents and a thorough inspection of the facility.

The exhibits shown at the end of this chapter are useful for all small quantity generators, regardless of the state in which they conduct their operations. These examples clearly demonstrate how detailed an inspection can be. Each item is marked with a corresponding reference to its applicable regulatory code, which the reader might find helpful.

The biggest single factor involved in an inspection is the regulatory agency doing the inspection. Enforcement by federal and state agencies varies widely, with such factors as the size of the generator, the experience of the inspector, the budget and available staff, and the severity of the violation being taken into consideration. While it is the objective of most regulatory agencies to inspect all generators, many small quantity generators have never been inspected. With the sheer volume of facilities, it is natural that larger generators would be more likely to be scrutinized by regulators. Regardless of the likelihood, it is advisable to perform an in-house environmental audit at least once a year to check compliance.

INSPECTING AGENCIES

Only two agencies are likely to inspect for compliance with hazardous waste regulations. These are the U.S. Environmental Protection Agency (EPA) regional office and the state agency responsible for administering hazardous waste regulations (see Appendix I). In states with EPA authority to administer regulations, inspections by EPA officials are infrequent, even for large generators. Since there are relatively few federal requirements for small quantity generators, but many state requirements, we will concentrate on state inspections.

WHAT STATE INSPECTORS LOOK FOR

Due to the differences in SQG definitions by the various states, the depth and scope of an inspection can vary considerably. In states where small quantity exemptions are extremely low, such as California and New Jersey, there is little difference between inspections for most generators. California does not have a small quantity exemption, and New Jersey conditionally exempts those generating less than 100 kg/month. Most inspectors concentrate on hazardous waste activity notification, manifest recordkeeping, waste labeling and storage, and accumulation times.

A typical inspection will frequently start with a request to see the generator's manifest file. While SQGs are not expected to have waste shipped off-site as frequently as larger generators, the inspector will automatically expand his inspection if there are no shipments or if the shipments are extremely infrequent. An inspector may feel he has the right to determine whether a "surplus" hazardous material is actually an undeclared waste. While standards vary considerably, inspectors are afforded a certain amount of latitude in making the determination of whether a material is actually a waste. The fact that this is legally the generator's decision in most cases may or may not be relevant. In most cases, showing some "good faith" efforts in having material shipped off-site for disposal will help the generator when questionable material is discovered on-site. In other words, inspectors want to see that a small quantity generator has made an effort to comply with regulations by showing some waste shipment activity.

For SQGs, accumulation beyond the allowable limits is probably the most common violation. Due to the problems of having small quantities of material transported off-site, this is not terribly surprising. Federal regulations have been amended to allow longer accumulation periods for small quantity generators, but many states have not adopted these limits. A second common violation is the storage of ignitable or reactive wastes less than 50 feet from the facility's property line.

While flagrant violations and obvious safety problems will quickly catch the eye of most inspectors, it is the more subtle requirements that are most often responsible for violations. Perhaps the most common violation of some state regulations, and one occasionally missed by inspectors, is the requirement that generators receive a letter of authorization from the designated treatment, storage and disposal facility that they are permitted to accept particular wastes from the generator. This is not a federal requirement, but many states have adopted it in one form or another.

Other common violations of various state laws include lack of contingency planning, lack of personnel training and training plans, improper segregation of containerized waste, containers leaking or without proper seals, and hazardous waste labels not clearly visible.

ENSURING SUCCESSFUL INSPECTIONS

There are two fairly effective ways to ensure a successful inspection with no serious violations. One is to perform an environmental audit as described in Chapter 7. The second is to develop a checklist of requirements that should be inspected or reviewed periodically, using this chapter as a guide, and to perform your own inspections.

Cooperation with an inspector by convincing him of an honest desire to comply with the regulations can make a great difference in an inspection. A noncooperative attitude can make the inspection much more of an adversary action. An inspector can help the generator to comply by making suggestions for improvements, and he should be able to answer specific questions on his interpretation of regulations that are difficult for the generator to understand. Since the inspector is the one who is most often called upon to interpret the regulations, it is important that his opinions be recognized.

Be sure to obtain the name and telephone number of your inspector in case you have specific questions later. A good relationship can be important to you and your facility, and a good cooperative attitude will undoubtedly help solve any problems that occur later. Last but not least, treat an inspection as a positive, learning experience, taking prompt corrective action when needed to ensure compliance in the future.

Following are reproductions of the RCRA Checklist for Small Quantity Generators of Hazardous Waste (Figure 1) and the State of Maryland Department of Health and Mental Hygiene Inspection Form (Figure 2).

RCRA Checklist for Small Quantity Generators of Hazardous Waste

R.O. USE

Inspection file No:

Name of Facility: _____

Address: _____

_____

Reviewer:

EPA Generator ID Number: _____

Date Reviewed:

Title: _____

Telephone Number: _____

Form "C"

The questions contained in this checklist apply to owners and operators who are small quantity generators (less than 1000 kg per month).

1. Has the facility identified all hazardous wastes generated on site in accordance with §262.11?

2. What types of waste are generated at the facility and the quantity of each per month?

_____ / _____   _____ / _____   _____ / _____

3. Does the facility treat or dispose of his hazardous waste in an on-site facility; or

Yes   No

ensure delivery to an off-site treatment, storage or disposal facility?

Yes   No

4. Does either the on-site (treatment, disposal) or off-site (treatment, storage or disposal) facility?

A. Have a Federal hazardous waste permit?

Yes   No

B. Have interim status?

Yes   No

C. Beneficially use or reuse, or legitimately recycle or reclaim hazardous waste?

Yes   No

D. Treat waste prior to beneficial use or reuse, or legitimate recycling or reclamations?

E. Have a State permit to manage industrial or municipal hazardous waste?

Yes   No

5. Please list name, address and EPA I.D. number for each facility where each waste is disposed.

Figure 1.   RCRA Checklist for Small Quantity Generators of Hazardous Waste.

-2-

6. Has the small quantity generator accumulated an amount
    of hazardous waste on-site, which is greater than?

    A. 1000 Kilograms?   Yes   No

    B. 1 Kilogram of acutely hazardous waste?   Yes   No

    C. 100 Kilograms of any residue, contaminated soil, water
        or debris from a spill of hazardous waste?   Yes   No

7. If so,

    A. Is the date upon which the accumulated amount in
        question 6 was reached clearly marked on the container?   Yes   No

    B. Has the hazardous waste been stored at the facility
        for greater than 90 days from the accumulation date
        in (A) above?   Yes   No

    C. Are the containers packaged, labeled and marked in
        accordance with DOT regulations?   Yes   No

    D. Is the hazardous waste stored in an on-site facility,
        which has interim status or a State/Federal hazardous
        waste permit?   Yes   No

Inspector's Name: _____

Title: _____

Agency: _____

Office Location: _____

Date of Inspection: _____

Inspector's Name: _____

Title: _____

Agency: _____ Office Location: _____

Date of Inspection: _____

Figure 1.   Continued.

RCRA CHECKLIST FOR INSPECTION OF GENERATORS

RO USE

Name of Facility:_____

Inspection file

Address:_____

No._____

_____

Reviewer_____

EPA Generator ID Number:_____

Date Reviewed:_____

Facility Inspection Representative:_____

Form "A"

Title:_____

Telephone Number:_____

Pert. Regs.
40 C.F.R.   1. Please provide a brief narrative explaining the
                type of work activity that occurs at the
                generator.

                _____

                _____

                _____

                _____

            2. Does the generator disposes of its wastes....

                A. On-site
                                    (Circle one)
                B. Off-Site

                Note: If on-site, then checklist for both a generator and
                       TSD facility must be completed if on-site more than
                       90 days.

Figure 1.   Continued.

2

3. Are 1000 kg (2200 Lbs) or more of hazardous   Yes No
    waste produced by the generator facility
    in a month?(If the amount is less than 1,000
    kg/month, then the facility qualifies as a
    small generator and Form C should be completed
    instead of Form A.)

4. What categories of hazardous wastes
    result from the generator's facility?

    A. Ignitable wastes   Yes No

    B. Reactive wastes   Yes No

    C. Corrosive wastes   Yes No

    D. EP Toxic wastes   Yes No

    E. RCRA Listed Wastes   Yes No

        Types _____   _____   _____

              _____   _____   _____

              _____   _____   _____

5. Is the generator presently...

    A. Treating hazardous waste?   Yes No

    B. Storing hazardous wastes longer   Yes No
        than 90 days?

    C. Disposing hazardous waste?   Yes No

    Note: If the generator performs any of the
           activities noted in Question 5, then
           the inspector must complete Form B,
           entitled "RCRA Checklist for inspection
           of hazardous waste treatment, storage
           and disposal facilities."

262.20   6. In a manifest system currently   Yes No
                in operation at the generator's
                facility so that offsite shipment
                of hazardous wastes can be tracked?

Figure 1. Continued.

3

7. Please inspect the generator's
    manifest for the following
    information

262.20   A. Is the **TSD** facility which receives   Yes No
                a generator's hazardous waste identi-
                fied by name, address, and EPA
                ID number?

262.20   B. Is an alternative facility designated   Yes No
                in case of an emergency? (Optional)

            C. Is a serialized manifest document number   Yes No
                included on the form?

262.21   D. Is the generator's name, address,   Yes No
                telephone number and EPA ID number
                included on the form?

            E. Is the name and identification number   Yes No
                of each transporter included on the form?

            F. Is a description of the generator's hazard-   Yes No
                ous waste to be treated, stored, or dis-
                posed included on the manifest?

            G. Is the quantify of each waste by units   Yes No
                of weight or volume and the type and
                number of containers loaded in the
                transport vehicle included on the
                manifest form?

            H. Is the following certification noted   Yes No
                on the generator's manifest form and
                is the certification acknowledged by
                the generator's signature.

                "This is to certify that the above-named
                materials are properly classified, described,
                packaged, marked, labeled and are in proper
                condition for transportation according to the
                available regulations of the DOT and EPA."

262.22   I. Are there adequate copies of the manifest   Yes No
                available for generator, transporter,
                and TSD's?

Figure 1.   Continued.

4

262.34(a)(1)	8.	Is all hazardous waste being shipped off-site by the generator within 90 days to a designated facility or placed in an on-site facility either of which has interim status or a Federal hazardous waste treatment, storage or disposal permit?	Yes No
262.34(a)(3)		A. Is the date accumulation of waste began clearly marked on each container?	Yes No
262.34(a)(2)		B. Are storage containers or tanks in good condition, i.e., no corrosion, leaking or structural deformations?	Yes No
		C. Starting at the time of initial accumulation are the storage containers	
262.34(a)(4)		1) Labeled	Yes No
262.34(a)(4)		2) Marked	Yes No
262.34(a)(2)		3) Packaged	Yes No
		as containing a particular hazardous waste in accordance with DOT regulations?	

Questions 9-15 apply to generators who accumulate wastes in a non-permitted facility.

265.16(a)	9.	Have facility personnel successfully completed a program of classroom training or on-the-job training in hazardous waste management procedures?	Yes No
265.16(d)	10.	Does the generator facility maintain a record of job titles for personnel that are involved with hazardous waste management and the name of the employee filling each job?	Yes No
265.16(d)(2)	11.	Does the generator facility have on record a written position description for each job title noted in Question #10?	Yes No
265.16(d)(3)	12.	Does the facility presently maintain a written description of the type and amount of introductory and continuing training for those employees noted in Question #10?	Yes No

Figure 1.   Continued.

5

265.32(a)	13.	Does the generator facility have installed the following equipment:	

| | | A. | An internal communications or alarm system capable of providing immediate emergency instructions to facility personnel if the hazardous waste storage area is threatened by fire or explosion? | Yes No |

| | | B. | A device at the scene of hazardous waste generator operations capable of summoning emergency assistance from Police, Fire departments, etc.? | Yes No |

| | | C. | Fire control equipment and an adequate supply of fire fighting water or fire supression chemicals? | Yes No |

265.35   14.   Does the generator facility have adequate   Yes No
aisle space to allow the unobstructed
movement of personnel and equipment
during emergencies?

265.50   15.   Does the facility have a contingency plan
which contains the following elements:

A.   Detailed description of emergency   Yes No
procedures facility personnel
will implement in response to
fires, explosions, or unplanned
releases of hazardous wates to
air, soil, and water?

265.52(c)   B.   A detailed description of arrange-   Yes No
ments formally agreed to by local
police, fire departments, and State
and local emergency teams to provide
assistance during emergency situations?

265.52(d)   C.   A listing of names, addresses, and   Yes No
phone numbers of the generator facility
emergency response coordinators?

Note: This listing should include names and phone numbers
of emergency coordinators available on twenty-four hour basis.

265.52(e)   D.   A list of appropriate emergency   Yes No
equipment necessary to cope with
emergencies at the generator facility?

Figure 1. Continued.

6

265.53   16. Has a copy of the contingency Plan been   Yes No
                submitted to local police, fire
                departments, hospitals, and emergency
                response teams that may be called on
                to provide emergency services.

           17. Please provide detailed explanation or
                comments on specific questions or problems
                encountered during the inspection. For
                instance, industry requests for exclusions
                from optional portions of the regulation or
                for clarification of specific RCRA rules and
                regulations and their applicability at the
                facility can be noted below or described in a
                separate memo attached to the inspector's
                checklist.

_____

_____

_____

_____

_____

Inspector's Name:_____

Title:_____

Agency:_____

Office
location:_____

Date of
Inspection:_____

Inspector's Name:_____

Title_____

Office
Location_____

Date of
Inspection:_____

Figure 1.   Continued.

State of Maryland
Department of Health and Mental Hygiene
Office of Environmental Programs
201 W. Preston St., Balto. MD 21201

YR	MO	DY

DHS Inspection Form
Generators/TSD Facilities

TIME

EPA ID Number

TELEPHONE

Owner/Operator _____ Facility Name _____

Address _____ Zip _____

Description of Work Activity _____

I. Generators

A. Description (10.51.03.01-.03)

1) Does the Facility generate or has it accumulated those quantities of hazardous waste described in 10.51.02.05 C.? ____Yes, ____No.

2) Has the facility obtained an EPA identification number? ____Yes, ____No.

3) Describe the amount of waste generated. (day, week or month)

4) Under which category is the waste(s)?
____Ignitable   ____Reactive   ____Corrosive
____EP Toxic   ____RCRA Listed

B. Manifest (10.51.03.04)

1) Is Maryland manifest system in operation for off-site shipment? ____Yes, ____No.

2) Is TSD Facility to receive DHS identified by ____Name, ____Address, ____EPA ID Number?

3) Is alternate facility identified? ____Yes, ____No.

4) Is generator identified by ____Name, ____Address, ____Telephone Number, ____MD/EPA ID Number?

5) Is each transporter identified by ____Name, ____EPA ID Number, ____Maryland Certification Number?

6) Is waste property described? ____Yes, ____No.

7) Is shipment date marked? ____Yes, ____No.

8) Is quantity of waste described by ____Unit of Weight, ____Volume?

9) Are containers to be loaded identified by ____Type, ____Number?

10) Is proper certification noted and signed by generator? ____Yes, ____No.

11) Are adequate copies available for operator, transporter and TSD? ____Yes, ____No.

C. Pre-Transport Requirements (10.51.03.05)

1) Is each container marked with date accumulation began? ____Yes, ____No. If yes, has any waste been stored over 90 days? ____Yes, ____No. How much _____

2) Are containers in good condition? ____Yes, ____No. If no, explain _____

3) Are containers properly labeled? ____Yes, ____No.

4) Does generator have approved emergency contingency plan? ____Yes, ____No.

D. Recordkeeping and Reporting (10.51.03.06)

1) Does the generator have: copies of all signed manifests from the previous three years? ____Yes, ____No; copies of each Annual Report and Exception Report? ____Yes, ____No.

2) Does the generator retain, for a period of three years, all wastes analyses? ____Yes, ____No.

3) Has the generator filed Exception Reports as required by 10.51.03.06 C? ____Yes, ____No.

II. Treatment, Storage, Disposal (TSD)

A. Site characterization (10.51.05.02)

1) Facility Type
____Thermal Treatment   ____Biological Treatment
____Recycling/Recovery   ____Land Treatment
____Waste Oil   ____Incineration
____Chemical Treatment   ____Landfill Operation
____Physical Treatment   ____Below Ground Tanks
____Open Pile   ____Other _____
____Surface Impoundment   _____
____Drums   _____
____Above Ground Tank(s)   _____

2) Does facility generate DHS? ____Yes, ____No.

3) Does facility have waste analysis plan? ____Yes, ____No. If yes, are the procedures of that plan being followed? ____Yes, ____No.

4) Can facility personnel identify DHS being handled? ____Yes, ____No.

5) Can facility personnel confirm that DHS received equal those on manifest form? ____Yes, ____No.

6) Is there a 24-Hour surveillance system to monitor active portion of facility? ____Yes, ____No.
If No, is there an artificial or natural boundary? ____Yes, ____No. Is there a means to control entry? ____Yes, ____No. Is there a restricted access sign posted? ____Yes, ____No.

7) Does facility have: ____emergency equipment inspection log, ____written schedule for inspections, ____security devices, operating & structural prevention equipment?

8) Have facility personnel completed classroom/on-site training? ____Yes, ____No.
Are records maintained of: ____Job titles/names of employees ____job descriptions, ____Type/amount of continuing training?

9) Are general requirements for Ignitable, Reactive or Incompatible Wastes as required in 10.51.05.02 H addressed? ____Yes, ____No.

B. Preparedness and Prevention (10.51.05.03)

1) Facility has the following equipment? ____Internal communication/alarm system for on-site personnel, ____device for summoning emergency assistance, ____adequate fire control equipment, water, & suppression chemicals, ____list of aforementioned equipment.

2) Does facility have adequate area for emergency movement? ____Yes, ____No.

C. Contingency Plan and Emergency Procedures (10.51.05.04)

1) Does facility have an approved contingency plan for:
____Personnel to implement emergency procedures to fire, explosions, and unplanned releases to air, soil and water?
____Responding emergency units to provide assistance during emergency situations?
____A list of emergency equipment needed to cope with situation?

2) Are emergency response coordinators listed by name, address, & phone number? ____Yes, ____No.

3) Is there an evacuation plan if recommended? ____Yes, ____No.

4) Are emergency coordinators available on twenty-four hour basis? ____Yes, ____No.

D. Manifest System, Recordkeeping, and Reporting (10.51.05.05)
Facility has a written operating record which contains the following information:

1) ____description & quantity of DHS received.

2) ____method & date of DHS treatment, storage, or disposal.

3) ____location & quantity at each DHS location in facility.

4) ____detailed records & results of waste analysis & treatability tests performed.

5) ____detailed operating summary reports.

6) ____description of emergency incidents that required implementation of contingency plan.

7) ____records & results of inspections of emergency equipment. TSD systems & hazardous waste areas.

8) Has facility retained, for at least 3 years, copies of all manifests? ____Yes, ____No.

Figure 2.   State of Maryland inspection form.

(2)

E. Groundwater Monitoring (10.51.05.06)
1) Has facility implemented a groundwater monitoring program? _____Yes, _____No, _____N/A.
2) Are samples from the groundwater monitoring system being analyzed according to the groundwater sampling and analyses plan? _____Yes, _____No.
3) Is this plan set up in accordance with 10.51.05.06 C? _____Yes, _____No.
4) Has groundwater quality assessment program been prepared? _____Yes, _____No.
5) Are proper groundwater sampling and analyses records kept? _____Yes, _____No.
6) Are the necessary reports on groundwater monitoring information being forwarded to the Secretary? _____Yes, _____No.
7) Do the reports match the facility records? _____Yes, _____No.

F. Closure, Post-closure, and Financial Requirement (10.51.05.07 & .08)
1) Does the facility have an approved closure plan that meets the financial requirements? _____Yes, _____No.
2) For surface impoundments, land treatment, and landfills, does the facility have an approved post-closure plan that meets the financial requirements? _____Yes, _____No.
3) Does facility maintain liability insurance? _____Yes, _____No.

G. Container Management (10.51.05.09)
1) Are all containers: (a)_____ in good condition, i.e., no signs of leakage, corrosion, or any other deterioration/deformation; (b)_____ lined or made of compatible material such that hazardous wastes placed into them will not result in reaction or corrosion; (c)_____ sealed during storage.
2) Are storage areas for hazardous waste containers inspected by owner/operator at least once a week? _____Yes, _____No.
3) Is an inspection log maintained? _____Yes, _____No.
4) Are containers holding ignitable or reactive waste located at least 50 feet from the facility's property line? _____Yes, _____No.
5) Are incompatible wastes placed in separate containers? _____Yes, _____No.
6) Are storage containers holding hazardous wastes which are incompatible with nearby materials stored in containers, tanks, piles, or surface impoundments separated by dikes, berms, walls, or other devices? _____Yes, _____No.

H. Tanks (10.51.05.10)
1) Are all tanks in good condition, i.e., no signs of leakage, corrosion, or any other deterioration: _____Yes, _____No.
2) Are uncovered tanks operated to ensure a minimum of two feet of freeboard? _____Yes, _____No.
If not, is tank equipped with a containment structure (e.g., dike or trench), a drainage control system, or a diversion structure (e.g., standby tank) with a capacity that equals or exceeds the volume of top 2 ft. of the tank? _____Yes, _____No.
3) Are tanks with continuous inflow of hazardous waste equipped with a means to stop this inflow (e.g., waste feed cut-off system or by-pass to a standby tank)? _____Yes, _____No.
4) Are waste analyses conducted or written documentation obtained before placing a substantially different hazardous waste into tank used for storage or treatment? _____Yes, _____No.
5) Are daily inspections conducted for discharge control equipment (e.g., by-pass systems, waste feed cut-off systems and drainage systems)? _____Yes, _____No.
6) Is data gathered from monitoring equipment (e.g., pressure and temperature gauges) at least once each operating day? _____Yes, _____No.
7) Is the level of waste in the tank checked at least once each operating day? _____Yes, _____No.
8) Is (are) the tank(s) inspected weekly to detect corrosion or leaking of fixtures or seams? _____Yes, _____No.
9) Are the results of these inspections recorded in an inspection log or summary? _____Yes _____No.
10) Are ignitable or reactive wastes stored in tanks? _____Yes, _____No. If yes:
a) Is the waste treated, rendered, or mixed before or immediately after placement in the tank so that the resulting waste, mixture, or dissolution of materials no longer meets the definition of ignitable or reactive wastes under Parts 261.21 or 261.23 of the RCRA Regulations? _____Yes, _____No.

b) Is waste stored or treated in such a way that it is protected from material or conditions which may cause the waste to ignite or react? _____Yes, _____No.
c) Is owner/operator of a facility which treats or stores ignitable or reactive wastes in covered tanks in compliance with the National Fire Protection Association's (NEPA's) buffer zone requirements for tanks contained in tables 2-1 through 2-6 of the "Flammable and Combustible Code—1977"? _____Yes, _____No.

I. Surface Impoundments (10.51.05.11)
1) Is two feet of freeboard maintained in the surface impoundment? _____Yes, _____No.
2) Do all earthen dikes have protective covers (e.g., grass, shale or rock) to minimize wind and water erosion and to preserve dike structural integrity? _____Yes, _____No.
3) Are waste analyses conducted or written documentation obtained before placing a substantially different hazardous waste into a surface impoundment used for storage or treatment? _____Yes, _____No.
4) Is the freeboard level inspected daily? _____Yes, _____No.
5) Is the surface impoundment, including dikes and vegetation, inspected weekly to detect leaks, deterioration, or failures in the impoundment? _____Yes, _____No.
6) Are the results of these inspections recorded in an inspection log or summary? _____Yes, _____No.
7) Are ignitable or reactive wastes stored in a surface impoundment? _____Yes, _____No. If yes:
a) Is the waste treated, rendered, or mixed before or immediately after placement in the impoundment so that the resulting waste, mixture or dissolution of material no longer meets the definition of ignitable or reactive waste under Parts 261.21 or 261.23 of the RCRA Regulations? _____Yes, _____No.
b) Are incompatible wastes segregated in separate surface impoundments so that spontaneous reactions are avoided? _____Yes, _____No.

J. Waste Pile (10.51.05.12)
1) Is wind dispersal of the pile controlled? _____Yes, _____No, _____Not Needed.
2) Are additions to the pile being analyzed prior to adding them to the pile? _____Yes, _____No.
3) Is hazardous waste leachate or runoff collected? _____Yes, _____No. Is the pile protected from precipitation and run-on? _____Yes, _____No.
4) Are ignitible or reactive wastes protected from materials or conditions that might cause it to ignite or react? _____Yes, _____No, _____N/A.
5) Are incompatible wastes hauled in a manner as to assure separation? _____Yes, _____No, _____N/A.

K. Land Treatment (10.51.05.13)
1) Will the use of land treatment result in the waste being less hazardous or non-hazardous? _____Yes, _____No.
2) Is run-on diverted away from the active portion of the facility? _____Yes, _____No. Is run-off from the active portion of the facility collected? _____Yes, _____No.
3) Has the proper waste analyses been peformed? _____Yes, _____No.
4) If food chain crops are to be grown on the active portion of the facility has the necessary documentation required been provided? _____Yes, _____No.
5) Has the owner/operator written and implemented an unsaturated zone monitoring plan? _____Yes, _____No.
6) Have the additional requirements for a closure and post-closure plan been addressed? _____Yes, _____No.
7) Are ignitable or reactive wastes immediately incorporated into the soil? _____Yes, _____No.
8) Are incompatible wastes hauled according to 10.51.05.13 I? _____Yes, _____No.

L. Landfills (10.51.05.14)
1) Is run-on diverted away from the facility's active portions? _____Yes, _____No.
2) Is run-off collected from the landfill's active portions? _____Yes, _____No.
3) Has a hazardous waste determination been made on the run-off? (Identification and Listing of Hazardous Waste) _____Yes, _____No.
4) Is the landfill managed so as to control wind dispersal? _____Yes, _____No.

Figure 2. Continued.

(3)

5) Are the following items maintained in the operating record: _____on a map, the exact location and dimensions, including depth, of each cell with respect to permanently surveyed benchmarks? _____contents of each cell and approximate location of each hazardous waste type within the cell?

6) Are bulk, non-containerized or waste containing free liquids placed in the landfill? _____Yes, _____No. If yes: _____ is a leachate collection system available to remove leachate?, and _____is the liquid stabilized or treated physically or chemically prior to disposal?

7) Are empty containers crushed flat or shredded before burial in the landfill? _____Yes, _____No.

8) Are containers holding liquid wastes (or waste containing free liquids placed in the landfill? _____Yes, _____No. If yes, describe containers on comments below.

9) Are ignitable or reactive wastes placed in a landfill? _____Yes, _____No. If yes: _____is the waste treated, rendered, or mixed before or immediately after placement in the landfill so that the resulting waste, mixture, or dissolution of material no longer meets the definition of ignitable or reactive waste? _____Are incompatible wastes segregated in different landfill cells?

M. Incinerator/Thermal Treatment (10.51.05.15 & .16)

1) Prior to burning waste not previously incinerated or thermally processed, does the operator conduct waste analysis for the following:
_____heating value of the waste;
_____halogen content and sulfur in the waste;
_____concentrations of lead and mercury unless documented data is available which show these elements not to be present?

2) Are instruments related to combustion and emission control monitored at least every 15 minutes? _____Yes, _____No.

3) Is the stack plume observed visually at least hourly for color and opacity? _____Yes, _____No, _____N/A.

4) Is the incinerator or thermal process and associated equipment inspected daily for leaks, spills and fugitive emissions? _____Yes, _____No.

5) Is all of the above information documented in the facility's operating record? _____Yes, _____No.

N. Chemical, Physical and Biological Treatment (10.51.05.17)

1) Are all treatment processes or equipment in good condition, i.e., no signs of leakage, corrosion or any other deterioration? _____Yes, _____No.

2) Are treatment processes or equipment with continuous inflow of hazardous waste equipped with a means to stop the inflow? (e.g., waste feed cutoff system or bypass system to a standby containment device) _____Yes, _____No.

3) Are waste analyses performed or written documentation obtained before placing a substantially different hazardous waste into treatment processes or equipment? _____Yes, _____No.

4) Is this information recorded in the facility's operating record? _____Yes, _____No.

5) Are daily inspections conducted for discharge control equipment (e.g., bypass systems, waste feed cutoff systems, drainage systems and pressure relief systems)? _____Yes, _____No.

6) Is data gathered from monitoring equipment (e.g., pressure and temperature gauges) daily? _____Yes, _____No.

7) Are construction materials of the treatment process or equipment and the immediate surrounding area inspected weekly for signs of leakage, corrosion or any other deterioration? _____Yes, _____No.

8) Are the results of these inspections recorded in an inspection log or summary? _____Yes, _____No.

9) Are ignitable or reactive wastes placed in a treatment process? _____Yes, _____No. If yes:
_____Are wastes treated, rendered, or mixed before or immediately after placement in the treatment process or equipment so that the resulting waste, mixture, or dissolution of material no longer meets the definition of ignitable or reactive wastes under Section 261.21 or 261.23 of the RCRA Regulations?
_____Are wastes treated in such a way that they are protected from any material or conditions which may cause the waste to ignite or react?

10) Are incompatible wastes kept from being placed in the same treatment process or equipment? _____Yes, _____No.

O. Permit Requirements (10.51.07)

1) Does the facility have a DHS permit for its activity? _____Yes, _____No.
If no, has the facility submitted an application for a DHS permit? _____Yes, _____No.

2) List any special Permit requirements that are not in full compliance. _____
_____
_____
_____
_____
_____
_____

Comments: _____
_____
_____
_____
_____
_____
_____
_____
_____
_____
_____
_____

Inspector's Name: _____ Title: _____

Facility Location: _____

Facility Rep. present during inspection: _____ Title: _____

Figure 2.   Continued.

9

Laboratory Waste Management

Laboratories represent a significant percentage of the total number of small quantity hazardous waste generators, though there are few comparisons that can be made between laboratories and most other small quantity generators. Laboratories can easily dispose of several thousand different chemicals in one year; most other small generators will dispose of only a few. Laboratories have the capability to treat many wastes to render them nonhazardous or less hazardous, yet disposal regulations do not necessarily allow laboratories to treat their own wastes. This chapter is not an attempt to provide a thorough treatise of on-site laboratory waste treatment and disposal, but instead is designed for those laboratories that are small quantity generators without available labor resources to commit to large-scale, on-site reduction or disposal of wastes, and that do not find that local regulations permit some freedom in this regard.

What types of waste are generated by laboratories? Essentially, the wastes comprise surplus, out-of-date, or otherwise unwanted and discarded chemical reagents, research preparations, waste solvents, analytical testing samples, and quality control samples. Various estimates exist on the number of chemical compounds known to man, but it is safe to say that there are potentially millions of different compounds that may be found in small quantities in the laboratory. The average manufacturing operation, on the other hand, generates only a few waste streams, but the quantities are much larger.

While excellent waste treatment and disposal techniques exist, the general failure of federal and state regulations to consider the special problems of laboratories have made many procedures illegal or open to various regulatory interpretations. While the U.S. Environmental Protection Agency (EPA) has defined and described a labpack (40 CFR 264.316), the most common packaging container for laboratory chemicals, this term only applies to the off-site landfill disposal of laboratory chemicals.

LABPACKS

A labpack is an overpack container, usually a steel or fiber drum, with small quantities of chemicals of the same hazard class packed inside with a packing material capable of cushioning bottles and absorbing spills from any liquid containers in the drum. Labpacks are sent off-site for processing, usually either by burial in a secure chemical landfill or by incineration. No provisions are provided for on-site treatment and disposal unless the laboratory can obtain a treatment, storage and disposal facility (TSDF) permit, a prohibitively expensive, time-consuming, and virtually impossible task for all but the largest and most powerful organizations. Laboratories are considered the same as other generators of hazardous waste despite the number of different waste streams, the size of those streams, and the technical expertise that is typically available in a chemical laboratory.

The U.S. Department of Transportation (DOT) allows the shipment of labpacks in an exception to normal packaging requirements. The procedure is detailed in 49 CFR 173.12, as follows:

> **(a) General.** Waste material meeting the hazard class definition of a flammable liquid, flammable solid, oxidizer, corrosive material, Poison B or ORM–A, B, C, and E are excepted from the specification packaging requirements of this subchapter if packaged in combination packagings in accordance with this section and transported for disposal or recovery by private or contract motor carrier by highway only. In addition, a generic proper shipping name from 172.101 may be used in place of specific chemical names, when two or more waste materials

in the same hazard class are packaged in the same outside packaging, providing the waste materials are chemically compatible.

(b) **Outside packagings.** The outside packaging must be a DOT specification metal or fiber drum. It may also be a polyethylene drum capable of withstanding: (1) The vibration and compression tests specified in 178.19–7 (c)(1) and (2), except the compression test value must be no less than 2400 pounds, and (2) a four-foot drop test as specified in 178.19–7 (a)(1).

(c) **Inside packagings.** The inside packagings must be either glass packagings not exceeding 1 gallon rated capacity, or metal or plastic packagings not exceeding a rated capacity of 5 gallons.

(d) **Additional packaging requirements.** The following additional requirements are applicable:
   (1) Each outside packaging may only contain one hazard class and the materials must be chemically compatible
   (2) Inside packagings of liquid must be surrounded by a compatible absorbent material capable of absorbing the total liquid contents; and
   (3) Gross weight may not exceed 450 pounds or the rated capacity of the drum, whichever is less.

(e) **Prohibited materials.** The following materials are not authorized under the provisions of this section: acrolein; bromine pentafluoride; bromine trifluoride; chloric acid; chlorine trifluoride; nitric acid, fuming; pyrophoric liquids; and sulfuric acid, fuming.

DISPOSAL IN MUNICIPAL SYSTEMS

As local regulations on municipal sewage and sanitary landfill disposal of waste vary considerably, laboratory waste managers are urged to investigate regulations thoroughly before putting chemicals down the drain or in the trash. Most sewer authorities limit the concentration of organic and other toxic materials that can be introduced into the system. The federal limit on toxic wastewaters from laboratories is 1% of the total annual wastewater volume from the entire facility. [40 CFR 261.3(2)(iv)(E)] Sewage treatment regulations for one urban county are summarized in Figure 1. From Figure 1 it should be obvious that disposal of laboratory wastes down the drain should be considered carefully. To reach a level of 1 mg/L for a 1-liter solution of any of these metals would involve dilution prior to discharge with several hundred thousand gallons of water.

Listed and characteristic hazardous wastes cannot be disposed of in sanitary landfills by regulated small quantity generators, and most sanitary landfills would undoubtedly restrict chemical wastes from homeowners if there were a good way to monitor such materials. A significant percentage of the sites on the NPL list, the federal list of chemical-contaminated sites needing cleanup, are sanitary landfills that in the past were unable or unwilling to control chemical dumping at their sites.

While the EPA and the various state agencies have done little to consider the differences between laboratories and other generators of waste, the U.S. DOT has, fortunately. The designation of a labpack as a proper packaging unit for many different hazard classes has greatly simplified the procedure for shipping laboratory chemicals off-site for disposal.

FACETS OF MANAGEMENT

Proper management of laboratory wastes involves many factors: program administration, the waste manager, training, inventory analysis and control, on-site treatment and disposal, storage, packaging, and labeling, off-site disposal method selection, selection of a transporter, recordkeeping and reporting requirements.

Program Administration

The best laboratory waste management programs are those that have the full support of the administration, both economically and philosophically. The higher the level of management support, the more likely that a program will be successful. Given the increasing likelihood of criminal as well as civil penalties for violations, the president and other executive officers of the organization should be actively involved in monitoring waste management activities. Input

Onondaga County, New York

The County of Onondaga Department of Drainage and Sanitation has published rules and regulations relating to the use of the public sewer system in the county. Prohibited discharges that may affect laboratory waste include the following:

3.01 (a) Any liquids, solids or gases which by reason of their nature or quantity are, or may be, sufficient either alone or by interaction with other substances to cause fire or explosion or be injurious in any other way to the treatment works or to the operation of the treatment works. . . . Prohibited materials include, but are not limited to, gasoline, kerosene, naphtha, benzene, toluene, xylene, ethers, alcohols, ketones, aldehydes, peroxides, chlorates, perchlorates, bromates, carbides, hydrides and sulfides. . . .

3.01 (c) Any wastewater having a pH less than 5.5 nor higher than 9.5 unless the treatment works are specifically designed to accommodate such wastewater . . .

3.01 (d) Any wastewater containing toxic pollutants in sufficient quantity, either singly or by interaction with other pollutants, to injure or interfere with any wastewater treatment process, constitute a hazard to humans or animals. . . .

3.01 (e) Any noxious or malodorous liquids, gases, or solids which either singly or by interaction with other wastes are sufficient to create a public nuisance or hazard to life or are sufficient to prevent entry into the sewers for their maintenance and repair.

3.06 Effluent Limitations and Concentrations . . .

Parameters	Daily Limit (mg/L)	Instantaneous Limit (mg/L)
Cadmium	2.0	3.0
Chromium, hexavalent	4.0	6.0
Chromium, total	8.0	12.0
Copper	5.0	7.5
Cyanide, total	2.0	3.0
Lead	1.0	1.5
Mercury	0.02	0.03
Nickel	5.0	7.5
Oil & Grease	100	150
Phenolic compounds	3.0	4.5
Silver	1.0	1.5
Zinc	5.0	7.5

County of Onondaga, Sept. 15, 1983.

Figure 1. Sewage treatment regulations for an urban county.

into a variety of different departments within the organization must be made to minimize liability and costs, the two major concerns of any waste management program, regardless of the size of the generator.

Many laboratories use a committee representing different departments to oversee the development and implementation of a waste management program. For instance, it is important to the waste management program to consider the purchase of laboratory chemicals, since the purchase of excessive quantities of chemicals can have a disastrous effect. Inventory control has the greatest overall effect on the cost of waste disposal. Security and safety are important parts of a program, and those with responsibility in these areas should certainly be involved. Representatives on the waste management committee should therefore include:

- Administration
- Laboratory supervisor
- Laboratory worker
- Safety manager
- Security department
- Purchasing

The Waste Manager

Any waste management program in a laboratory setting that does not have one person with overall responsibility will have difficulty succeeding. In most small laboratories this will not be a full-time position, but someone must control what goes into waste storage and how it is handled. This is particularly important in respect to waste segregation and the management and disposal of unknown materials. The waste manager should have training in hazardous materials management, both to provide a good foundation for the program and to meet federal and state requirements for training of those employees who handle chemical wastes. The waste manager can, in turn, be responsible for the training of other individuals who are involved in handling of waste.

One key function of the waste manager is control of the waste storage area. This area should not be accessible to all employees from both a safety and an inventory control standpoint. The waste manager should know at all times what materials are stored in the area and be able to verify that they are all appropriately labeled and stored. An accurate inventory should be kept, particularly in the case of compatible wastes that are bulked into larger containers for disposal. The addition of incompatible materials can result in both a safety hazard and higher disposal costs.

The following list summarizes the functions of the waste manager:

- Coordination of waste management program with administration
- Provision of training
- Control of access to storage
- Coordination of waste collection
- Supervision of packaging of wastes
- Supervision of selection of disposal options and contractors
- Implementation of waste minimization procedures

Training

Training will remain a nebulous requirement of federal waste management regulations until the EPA develops definitive standards for the training of personnel who handle waste. For those who are involved in the handling of laboratory waste, the delay in establishing standards is both a blessing and a curse. Based on previously promulgated regulations, the special situations faced by laboratories will not be addressed, resulting in a training outline that will be of little use in addressing the handling of laboratory waste. This may be a blessing because commercial training courses especially designed for laboratories have slowly become available. In addition to the commercial courses, a good training course can be developed by a laboratory with references that are readily available. Many of the appropriate procedures for handling hazardous wastes are identical to those common in the laboratory; good safety practices in the handling of laboratory chemicals in everyday use are the best basis.

Training should focus on compatibility of chemicals, safe storage and handling, proper labeling and segregation of wastes, and emergency response procedures. Following is a suggested training outline.

    I. Storage of hazardous waste
        A. Compatibility with container and environment
        B. Storage system design and implementation
        C. Ventilation
        D. Regulations regarding storage
        E. Labeling

II. Handling of hazardous waste
  A. Safety procedures in the laboratory
  B. Personnel protection
    1. Use of safety and emergency response equipment
    2. Use of personal safety gear
  C. Packaging waste materials
    1. Segregation guidelines
    2. Repackaging or bulking procedures

III. Disposal of hazardous waste
  A. Waste minimization
  B. On-site treatment and disposal methods
    1. Regulatory interpretations
  C. Off-site disposal
    1. Appropriate disposal method selection
    2. Selection of transporter and disposal facilities

IV. Regulatory compliance
  A. Evaluation of federal, state, and local regulations
    1. Recordkeeping and reporting
    2. Storage limits
  B. Labeling, packaging, and shipping requirements

Inventory Analysis and Control

Perhaps no other small quantity generators have more opportunity to control their inventory and reduce disposal costs than laboratories. Analytical and bench chemistry procedures developed over the last 20 years have made the use of small quantities of chemicals more practical in the laboratory. As microgram and gram quantities of many laboratory reagents are all that is required in many applications, the purchase of smaller quantities makes sense for a number of reasons. The purchase cost of the reagent is lower, glassware and other equipment may be smaller and less expensive, and waste disposal quantities are decreased. This "less is better" philosophy is important in any effort to reduce the costs of laboratory waste disposal.[1] It is unfortunate that only a few reagent distributors have begun to sell chemicals in containers smaller than the traditional 500-gram or liter quantities. The overwhelming percentage of laboratory waste disposed of off-site is unused or outdated reagents that were purchased in excessive quantities. Economy clearly dictates the purchase of smaller quantities whenever practical.[1,2]

An additional problem is poor control of this excessive inventory. If laboratory reagent distribution is decentralized, there is a strong possibility of duplicate purchasing. Sooner or later, someone will purchase a chemical that is already in stock somewhere else in the facility. For this reason, a centralized storage and inventory control system is most desirable. The additional purchase cost for unneeded chemicals, added to the disposal cost if the material is not used, will substantially increase the overall cash outlay for the material. Since disposal costs can frequently equal or exceed the purchase price, purchasing excessive quantities is akin to double jeopardy. To pay twice for a chemical that was not required in the first place does not make any sense.

How can a laboratory best analyze its reagent inventory requirements and prevent excessive purchasing of chemicals? The most effective way is to establish an inventory control system based on centralized storage. Historical data can be developed over time to estimate requirements. It is imperative to control what chemicals are removed from storage, preferably by a sign-out policy. Different types of laboratories obviously will vary in their abilities to estimate requirements in advance. A research and development (R & D) laboratory will have more difficulty than a testing or an academic laboratory because requirements are less defined. R&D laboratories are also more likely to need quick delivery of a chemical that might not be in stock. For this reason, R&D labs would normally have a larger inventory of chemicals, but could probably purchase those chemicals in smaller quantities. It is this concern over the packaging sizes that is the key to controlling costs once an inventory system has been implemented. Given the trend in analytical work to use smaller and smaller quantities, it would appear to be only a matter of time before the number of companies selling reagents in gram and microgram quantities will increase.

A number of companies with large R&D expenditures have developed sophisticated computer inventory systems for their chemical storerooms. A simple inventory system is not necessarily a large expense. Academic laboratories in particular may already have access to a computer center with sufficient resources to manage a chemical inventory, and there may be an additional cost advantage in combining inventory with hazard communication information for each material. State or local "Right to Know" legislation may require annual reporting or recordkeeping that can be added to a database.

On-Site Treatment and Disposal

Along with the reduction of waste by minimizing purchases and controlling inventory, there are a number of ways to reduce the quantity of waste shipped off-site for disposal. The most effective way is to treat the waste to make it less hazardous.

Treatment in many cases can render the waste non-hazardous, which may make it suitable for disposal in a municipal waste system. Although some of the regulations affecting a laboratory's ability to treat its own waste are unclear, a general interpretation is that if the end-product of a process or experiment is treated as part of the process (in the same location), then the waste can be treated. More liberal interpretations are based solely on the argument that a laboratory should have the technical expertise to treat its own wastes safely and effectively. By federal regulation, a conditionally exempt generator (less than 100 kg/month) may treat its waste on-site without a permit.

The most common method utilized to dispose of waste on-site is to pour it down the drain. This can be appropriate for a number of biodegradable materials. According to 40 CFR 261.3 (iv)(A), the Resource Conservation and Recovery Act may not define laboratory wastewater as hazardous waste if it is not otherwise excluded by reason of listing or characteristics. Such a waste is subject to Clean Water Act standards, excerpted as follows:

> [H]owever, the following mixtures of solid wastes and hazardous wastes listed in Subpart D are not hazardous wastes: . . . wastewater resulting from laboratory operations containing toxic (T) wastes listed in Subpart D, provided that the annualized average flow of laboratory wastewater does not exceed one percent of total wastewater flow into the headworks of the facility's wastewater treatment or pre-treatment system, or provided the wastes combined annualized average concentration does not exceed one part per million in the headworks of the facility's wastewater treatment or pre-treatment facility. Toxic (T) wastes used in laboratories that are demonstrated not to be discharged to wastewater are not to be included in this calculation.

The federal standard does not have jurisdiction over local sewer authorities, however. Municipal sewer authorities have published limits on a number of chemicals that are difficult to treat or that cannot be treated. The allowable concentrations vary considerably, but organic solvents, heavy metals, and pesticides are almost guaranteed to allow either concentrations in the low parts-per-million range or no detectable concentration at all. Whoever first said "dilution is not the solution to pollution" may have had the right idea. Diluting 1 liter of benzene to a 1-ppm concentration would require the addition of over 250,000 gallons of water!

Storage, Packaging, and Labeling

Storage

The storage of laboratory waste is partly a factor of the disposal methods to be utilized for the wastes. Waste materials from the laboratory are generally divided into two groups: (1) discarded chemical reagents and samples, and (2) waste solvents. We will first consider the storage of reagents and samples.

As chemical reagents and samples are not necessarily waste products until they are declared as such, these materials will not usually be kept in a central waste storage until they are declared as waste. One advantage of laboratories in the overall regulatory scheme is that most materials are not process wastes and, hence, are not obviously waste in the same respect as those materials generated by manufacturers and other waste generators. With limitations on accumulation time being somewhat restrictive, it may be advantageous to refrain from moving chemicals that could be considered surplus until shortly before a waste shipment from the facility.

Once a waste shipment is planned, surplus reagents and samples to be discarded should be inventoried and moved to a central storage area. This will help reduce the cost of having the material removed, since an inventory can be utilized to solicit competitive bids, and central storage will reduce the contracted labor time involved in the packaging for shipment.

Laboratories that generate waste solvents should consider whether the quantities generated warrant combining of materials prior to off-site disposal. Generally, if the quantity generated equals or exceeds 50 gallons in the period of time between waste pickups, it is advisable to consider bulking solvents. The compatibility of solvents and the disposal method to be utilized should be considered. It is normally not advisable to bulk flammable and nonflammable solvents together, since flammable solvents can be incinerated easily and inexpensively. The addition of halogenated solvents (those with chlorine, bromine, etc.) will make the disposal more expensive. These materials should be segregated, and the 50-gallon determination should be made separately. Bulking of solvents should be performed in a well-ventilated area, and personnel protective equipment should be utilized. These materials should be bulked directly into appropriate DOT specification containers prior to shipment. Storage should be in a well-ventilated area that is not exposed to high temperatures, and all solvent containers should be properly grounded.

Packaging

As previously mentioned, most waste containers of 5 gallons or less should be disposed of in labpacks. Regardless of the disposal method to be utilized, labpacks are the most appropriate and economical packaging units for small quantities of chemicals. Unless the generator is dealing directly with a disposal facility for waste approvals, it is not advisable for the generator to pack labpacks themselves. Most transporters and waste management companies want to pack drums themselves to ensure that all materials are compatible, properly labeled, and acceptable to the disposal facility. There is a great deal of liability associated with proper packaging, and few small generators have the expertise and time to learn the nuances of transportation regulations. Training of a chemist to labpack drums properly can easily take six months to a year. For this reason, it is best to let the contractor pack drums for shipment. Obviously, the selection of a contractor with experience is most important.

The packaging of bulkable solvents for shipment involves several considerations. The amount of space available for storage is important, since many small laboratories may not be able to store 55-gallon drums. For these laboratories, storage in smaller containers (5–30 gallon) may be appropriate. Also important is the use of qualified people to handle the bulking and the proper use of safety equipment and procedures.

Labeling

Labels for surplus or waste chemicals (once they have been declared a waste) in storage should include the proper name of the material (including all constituents, if a mixture), the hazard class, and date that accumulation began. Labels should be affixed properly, and should take into account any environmental conditions that will affect their integrity. Weather conditions, for those materials stored outside, should be taken into consideration. Many labels and markings cannot withstand rain, sunlight, or other adverse weather conditions.

Off-Site Disposal Method Selection

There are three basic off-site disposal methods for laboratory wastes: incineration, secure chemical landfilling, and "processing," which is a combination of chemical treatment, incineration, and landfilling. Choosing the appropriate disposal method for a generator's materials is extremely important in terms of both long-term and short-term considerations.

Incineration, for example, limits or virtually eliminates long-term liability to a generator. Once waste materials have been incinerated, there is nothing to be tracked back to a generator. This is generally the most expensive disposal option, and capacity at incineration facilities is becoming more and more in demand every year. With the bans on landfilling of many materials, incineration has become the best available option for the disposal of many waste materials.

Secure chemical landfilling has long been the most popular disposal method for laboratory wastes. The EPA specifically allows the landfilling of labpacks, though regulations and disposal facility policies limit the types of materials that can be landfilled in labpacks. Reactive and explosive materials, for example, are generally prohibited from landfill disposal. In many regions of the United States, landfilling is the only available disposal option for labpacks. The western United States, for example, has no commercial incinerators, and generators that want their materials incinerated must transport them great distances. Landfilling is essentially land storage, in that the waste materials do not disappear, and drums always can be traced back to the generator. There is a certain amount of long-term liability associated with the landfilling of hazardous wastes. At the moment, landfilling is the least expensive disposal option for hazardous wastes.

The processing of labpacks at a permitted TSDF is a relatively new way of handling the disposal of laboratory wastes. Essentially, processing is based on the best available disposal option for each chemical in a labpack. Labpacks are packed according to DOT regulations at the generator's site, then unpacked at a processing facility. Liquid waste materials that can be bulked for incineration are separated, as are acids and bases that are suitable for chemical treatment. Other materials may be either bulked for landfill disposal or labpacked again for either landfill or incineration. Liability is generally limited by labpack processing, but again the selection of a transporter or waste management company is extremely important (see Chapter 7).

Selection of a disposal method will normally depend on such factors as distance to disposal sites, size of the laboratory and quantities of waste generated, and financial resources of the generator. If a disposal facility goes bankrupt or develops regulatory problems, federal and state regulations may allow costs for cleanup of the site to revert to generators. In this event, those generators with "deep pockets" are usually the first generators asked (or ordered) to contribute to costs. In this event, a generator will effectively pay twice for the disposal of the same waste materials.

References

1. *Less is Better, Laboratory Chemical Management for Waste Reduction,* (Washington, DC: American Chemical Society, Task Force on RCRA, 1985).
2. Phifer, L. H., and Mathews, C. "A Small Quantity Approach to Laboratory Economy and Safety," *Am. Lab.* (August 1978).

10

Disposal Options

This chapter assumes that small quantity generators of hazardous waste do not have the means to permit their own on-site treatment or disposal facilities and must ship wastes off-site for disposal. The cost of permitting and operating a treatment, storage and disposal facility are prohibitive for all but the largest companies, and geographical considerations may prohibit certain disposal options in a particular region. Small quantity generators (SQGs) will normally have more than one disposal option available, however, and should carefully consider all of the ramifications of each suitable method. This decision should not normally be left up to a waste management contractor, although advice from a good consultant may be appropriate. These decisions may be contingent on available transportation options, quantities and types of waste generated, capacity of facilities, and permit requirements.

There are many technologies for the disposal of hazardous wastes, and more and more research is going into the development of additional options. While different waste streams may be ideally suited to particular disposal methods, these options may not always be practical due to a number of factors. The location of the generator in relation to disposal facilities is important. Costs for transportation of small quantities are not substantially lower than for larger quantities because the cost of putting a truck on the road remains roughly the same. To keep transportation costs reasonable, it is necessary in most cases for materials from more than one generator to be combined on a truck. The concept of a "milk run" is not new; most transporters of hazardous waste are used to combining loads. This does, however, put more of a burden on the generator to verify proper handling of paperwork, labeling, placarding, and loading procedures.

All hazardous waste disposal methods have drawbacks designed to make the selection of an appropriate method confusing. For this reason, reclamation and recycling are the most attractive options when they are available. Unfortunately, most small quantity generators do not have large enough quantities of anything to make these methods economical. With the possible exception of certain precious metals and some solvents, SQGs are forced to dispose of waste materials without gaining anything in return.

Following is an outline of the various disposal options. Each method and its pros and cons will be discussed. While this may appear to be an overwhelming list, there are rarely more than one or two disposal methods suitable for any particular waste. Unfortunately, many disposal methods result in residual materials that must still be disposed of properly. As some of these methods are totally unsuitable for small quantities of waste, we will concentrate on those methods that are most often utilized by small quantity generators.

I. Thermal Treatment

    A. Incineration. Incineration currently enjoys the reputation of being the most favorable disposal method for hazardous waste. Different types of incinerators are designed to handle different types of waste.

        1. Single-chamber liquid injection, for instance, is suitable only for liquid materials. It is a relatively inexpensive disposal method that can destroy a wide range of materials, including solvents and polychlorinated biphenyls.

        2. Cement kilns also are utilized extensively to burn solvents, with energy recovered from the process then used in the manufacture of cement.

        Regulatory changes and a greater concern over long-term liability have led to many solid materials being incinerated. Bans on the landfill of many wastes have resulted in significantly increased demand on commercial incinerators.

3. **Rotary kilns and multiple hearth furnaces** are examples of incineration methods suitable for many liquids and solid materials.
4. **Fluidized bed**
5. **Ocean vessel incineration**

The chief disadvantage of incineration is its cost. Due to the high expense in getting incinerators permitted, the high demand, and frequently high operating costs, incineration is generally the most expensive disposal option. Due to the high demand, many commercial incinerators have lead times approaching four to six months, and gaining approvals for disposal may also be a lengthy process. Since there are so few commercially available incinerators, transportation can represent a significant cost. The western United States, in particular, has a shortage of all types of thermal treatment facilities.

Incineration offers perhaps the lowest long-term liability, since waste materials that have been incinerated are essentially destroyed. The remaining ash is not trackable back to a generator. Another advantage is the wide range of materials that are suitable for incineration. Being able to send all of a generator's waste to one facility can reduce transportation and paperwork costs as well.

B. **Detonation.** Detonation is an accepted disposal method for many hazardous wastes, particularly explosive and highly reactive materials. It is utilized primarily for organic materials, which undergo molecular separation at the extremely high temperatures that are generated. Detonation is usually performed at the generator's site, which eliminates the need to transport unstable materials. Detonation is often performed underground, with a small booster charge initiating the blast.

Examples of materials suitable for detonation are picric acid, organic peroxides, peroxidized ethers, and many nitro compounds. Detonations must be carried out by specially trained personnel with blaster licenses. A disposal permit must be applied for on a case-by-case basis, and specialized safety equipment is involved. Only small quantities can be detonated at a time, and even with the controlled nature of detonations today there are limitations on where this operation can be performed. Urban locations, for instance, are usually not suitable. The federal regulations currently specify, for instance, that there must be a buffer zone of 670 feet from the property of others. Detonation can be extremely expensive, yet it is the most suitable disposal method for certain materials, particularly those that are unsafe to transport. Municipal bomb squads have traditionally offered this service, though permit requirements and liability issues have minimized this practice. Military bomb squads will not handle commercial work except in extreme cases.

C. **Pyrolysis.** Pyrolysis is similar to incineration, but is based on thermal destruction without the presence of oxygen. This reduces the formation of oxides, which create problems for conventional incineration facilities. Unfortunately, pyrolysis units are rare, the procedures are still being developed, and the economics do not dictate heavy usage at this time.

D. **Open burning.** Regulatory restraints have virtually eliminated open burning as a disposal option. Possible air and land pollution are cited as reasons. Some materials are disposed of illegally in this way, and some states may allow this method on a case-by-case basis for certain materials, such as waste explosives.

In summary, thermal treatment methods are suitable for nearly all wastes, though they can be expensive. The reduction of dependency on land disposal makes incineration and other thermal treatment environmentally attractive options, with lower long-term liability than land disposal methods.

II. **Chemical treatment.** Chemical treatment of waste can make it less hazardous and more amenable to land disposal or discharge to a sewage treatment plant. Essentially, it involves changing the chemical structure of the waste material.

The advantages of chemical treatment as a disposal method include a relatively low cost compared to incineration, yet low long-term liability. When a material is treated, it loses its identity, thus reducing the generator's exposure. Treatment can sometimes create a useful by-product, but more often than not it involves the creation of another less hazardous waste.

Chemical treatment plants usually process large quantities of material at one time, with waste from many generators being added together in a specified mix. Most plants are designed for only one treatment method, and liquids are generally more amenable to chemical treatment.

A. **Acid/base neutralization.** This is the most common type of chemical treatment. As corrosivity is a characteristic of many wastes and is frequently a function of pH, adjusting the acidity or alkalinity of a material to a neutral range can frequently eliminate the specific hazard associated with a waste. This does not result in the actual destruction of a waste, but will render it less hazardous and suitable for additional treatment and eventual safe disposal in the environment. Acidic and alkaline solutions are suitable for this treatment method.

B. **Carbon absorption.** Carbon absorption can be used to extract certain solvents from aqueous solutions. This is accomplished by moving the waste material through a stationary bed of activated carbon.

C. **Ion exchange.** The removal of negatively or positively charged particles is useful for extracting many metals from aqueous solutions. Precious metals are frequently reclaimed in this manner from plating and other metal finishing operations.

D. **Oxidation/reduction.** Frequently referred to as a redox process, oxidation/reduction is a common treatment method for cyanides and other reactive materials. A fairly wide range of other materials, both organic and inorganic, are also suitable for redox. The procedure involves breaking chemical bonds by moving electrons from one material to another. This can result in a less hazardous material in many cases.

E. **Precipitation/clarification.** Precipitation removes suspended or dissolved materials from a process, usually by the addition of an alkaline material such as lime. An insoluble solid forms which can be dewatered and landfilled as deregulated or less hazardous material. The remaining liquid is clarified and processed further in a wastewater treatment plant. This process is used in many industrial plants, particularly those that generate acidified metal-bearing wastes.

III. **Biological treatment.** The argument can be made that biological treatment processes are the wave of the future. Advances in genetic engineering promise technology for destroying a wide variety of wastes. The most simple methods of biological treatment have been utilized for many years. These processes all use microorganisms to decompose waste materials.

A. **Microorganisms.** Simply put, many microorganisms eat waste products. As digestion takes place, highly complex materials can be detoxified by breaking them down into natural materials. Municipal sewage treatment plants utilize microorganisms to treat sludge, and large industries frequently treat wastes on-site in large lagoons or pits. Concerns over leaching of materials have resulted in additional lining and monitoring requirements for lagoons, and few small generators need to process such large quantities anyway. Nonetheless, it will be interesting to see where research in this area will lead; the potential is enormous.

B. **Microbial breakdown.** Activated sludge, aeration, anaerobic digestion, and composting are all examples of microbial breakdown of waste. While these methods are all based on the processing of relatively large quantities of waste, all are environmentally sound methods for handling specific waste streams.

C. **Landfarming.** Landfarming has been utilized for many years to handle the disposal of large quantities of waste. The process basically depends on the environment to detoxify waste materials, and has many limitations. Landfarming requires the use of large areas of land, and this land cannot be utilized for agricultural purposes for many years. It is only suitable for certain materials, and can result in pollution of other land if not carefully monitored.

Biological treatment shows great promise for the disposal of a wide variety of different wastes. The technologies that are now being developed may eventually reduce or eliminate the need for less environmentally sound methods.

IV. **Land disposal.** Land disposal represents long-term storage in or above the ground. It is by far the most popular method of disposal historically since low cost and large capacity has in the past meant almost automatic selection of this disposal option. Environmental concerns have resulted in recent (and future) bans on the land disposal of many materials.

It is a stated goal of the legislative and regulatory community to completely eliminate the landfilling of hazardous wastes within the next decade. Whether or not alternative disposal methods and capacity exists currently, it seems evident that the disadvantages of land disposal outweigh the advantages in the long run.

Landfilling results in essentially eternal liability for a waste material. Since metal drums have been utilized extensively for the land burial of wastes, it is relatively easy to track material back to the generator. Permits require facility operators to keep track (usually with a grid record) of where specific material is stored, and there is no guarantee that any corporate concern can offer "perpetual care." It is assumed that eventually all landfills leak, and collection of leachate is necessary to prevent pollution of the environment. Regulatory requirements for land disposal facilities have increased dramatically since 1980, and costs for land disposal have also increased. This has closed the distance between the cost of landfilling and alternative methods that offer lower liability to the generator.

A. **Secure chemical landfilling.** Landfilling of waste materials in a designated "secure" chemical landfill is traditionally the catch-all disposal method. New, modern facilities have sophisticated liners and leachate collection systems that make them far safer than previous facilities. The cost of using these facilities remains lower than the cost of incineration, and capacity is basically available in all areas of the country. Nonetheless, many facilities are approaching capacity, and siting difficulties will certainly make it unlikely that new facilities will be built. There will always be a requirement for landfilling in one sense or another, though. Incineration and treatment technologies can still generate nonhazardous or less hazardous materials that need to be landfilled. The long range objective, therefore, is to eliminate the landfilling of hazardous materials to increase the available capacity for less hazardous materials.

B. **Deepwell injection.** Deepwell injection is based on drilling into the earth below the groundwater table, and then depositing hazardous wastes in the ground where they ostensibly cannot pollute natural resources. There are many unanswered questions concerning this method, especially concerning the potential for leakage in old facilities. Regulatory pressure has also resulted in a decrease in available facilities. Wastes from the petroleum industry have traditionally been disposed of by deepwell injection, and it is a relatively inexpensive option for large quantities of waste which has less potential for contamination of groundwater than other land disposal methods.

C. **Surface impoundment.** Almost a treatment method, surface impoundment is based on reducing large quantities of waste by evaporation. It is suitable for many aqueous solutions but, again, is not suitable for use by small generators. These facilities are typically on a generator's site, and regulatory concerns have resulted in strict requirements for lining and monitoring of lagoons, ponds, and impoundment pits.

DISPOSAL METHOD SELECTION

It is necessary to develop a priority approach to disposal method and site selection. Taking into account concern that the organization might have about long-term liability and balancing the short-term cost will result in the best decision-making process. Waste management companies and transporters will almost always handle the decisionmaking and approval process for a small generator, but the decision properly rests in the hands of the generator. Review the options carefully with potential contractors, and do not hesitate to suggest and investigate what appears to be the appropriate option. Many contractors will press a particular disposal option, but the generator retains the bulk of the liability over site and method selection.

11

Trends in Hazardous Waste Management

POLITICAL ASPECTS

Ever since protection of the environment has become an important media consideration, the public has taken the issue of proper waste management to heart. Regrettably, this has sometimes caused legislation and regulations to be rushed through without complete thought being given to the practicality of all of the policies established. Whether or not environmental protection became a politically advantageous issue in the 1970s, there is little doubt that politicians found it a popular platform. Still, the public has rightfully become concerned and knowledgeable enough about the condition of the environment to encourage the reduction of dependency on land disposal of hazardous wastes. While land disposal continues to be the cheapest of disposal technologies from an initial cost standpoint, there are implications beyond initial cost that must be considered.

LAND DISPOSAL: AN INTERIM STEP IN DISPOSAL TECHNOLOGY

As land disposal is and always will be simply land storage, it represents nothing more than a stopgap measure for proper waste management. Just as the disposal of nonhazardous wastes has begun to move toward such technologies as trash-to-steam to solve space problems, the hazardous waste management industry has moved towards thermal, chemical, and biological treatment methods as the answers to disposal problems. The days of siting huge land disposal facilities are over. In the disposal of nonhazardous as well as hazardous wastes, the public has shown it will not stand for landfills as the long-term solution to our disposal problems. While alternative technologies still have not been fully developed for many waste materials (particularly heavy metals), there is little doubt that eventually methods can be developed.

As Congress has essentially mandated that land disposal of untreated hazardous wastes be phased out by 1990, the industry fight to extend deadlines has been an uphill struggle. A number of landfill bans are already in effect, including most liquids, halogenated (chlorinated) organic materials, and solutions containing heavy metals. Whether sufficient alternate disposal technologies are available or not, there is no doubt that legislators and regulators are not likely to allow extensions much beyond current deadlines. What this will do to disposal costs is an important question. Those companies that have been successful in developing new technologies and getting facilities sited and permitted have a distinct advantage as deadlines approach.

MOBILE AND ON-SITE TREATMENT ADVANCES

The Environmental Protection Agency (EPA) is moving rapidly toward streamlining the permitting of mobile treatment units on a statewide basis instead of the previous site-specific standard. While many of these technologies are not new, there has been a significant push in the last few years to allow some treatment processes to be mounted on vehicles and mobilized; the EPA itself has pioneered mobile incineration at its dioxin decontamination site in Times Beach, Missouri. There are numerous advantages to mobile treatment units, particularly for the neutralization or destruction of large quantities of contaminated soil. Because of the high cost of hazardous waste transportation, moving large quantities of contaminated soil off-site can be prohibitively expensive. Treating wastes in situ, or, literally, where they sit, can substantially reduce the cost of material handling and transportation.

With many regions of the country lacking disposal facilities, either through citizen pressure or geological or ecological reasons, incredibly high costs for hazardous waste transportation present a nightmare for small quantity generators in particular. A recent trend toward the requirement of large cash bonds for state permitting has resulted in many small haulers being frozen out of certain states. One answer to these transportation problems is treatment on-site or in mobile units stationed at a generator's site. There is no logical reason why small quantity generators could not take advantage of these potential solutions, as they traditionally bear a higher per-unit cost for transportation than larger generators.

RESEARCH FUNDING

That hazardous waste management is still a relatively new industry and that traditional disposal methods have been inexpensive has provided little incentive in the past for large expenditures on research into new technologies. As disposal costs have risen dramatically in the last decade, however, there has been a new push for development of new technologies. Public pressure to discourage landfilling of hazardous wastes has forced the hazardous waste industry and environmental agencies to increase research efforts. As a result, there are numerous opportunities for funding available for new research. The emphasis here is on research into new alternate disposal technologies. The U.S. EPA provides funding to a number of research organizations, both in the academic and industrial sectors.

Colleges and universities have generally been slow to offer courses in hazardous waste management. A few exceptions, such as Texas A&M, Louisiana State University, and the University of Pittsburgh, have offered courses and research opportunities through related institutes as well as existing technical departments. Interestingly enough, however, no colleges or universities offer a major in hazardous waste studies. The first to offer such a program is Findlay College in Findlay, Ohio. This program has been initiated largely through the efforts of a local hazardous waste management firm. An obvious value of such a partnership between business and academia is the opportunity for waste management firms to hire graduates who have some training in the field. This has traditionally been a shortcoming of the hazardous waste industry; virtually all employees must undergo extensive training due to the lack of college-level courses in the field.

ADDITIONAL WASTE LISTINGS AND DELISTINGS

While the toxicology of hazardous waste has been largely an inexact science, there is a continuing drive to require risk assessment as part of any decisionmaking on hazardous waste listing. As more data is accumulated, it is certain that major changes in both the definition of Extraction Procedure (EP) toxic (see Chapter 2) and other possible characteristics of toxicity will result in the addition of hundreds or thousands of compounds listed.

Currently, many of the most toxic chemicals known to man are not listed or do not meet the existing definition of hazardous waste. Some examples are highly carcinogenic compounds such as pyrene and nitrosomethylethylamine (although similar compounds are listed). Pesticides such as 2,2-dichloropropionic acid and piperonyl butoxide are also not listed. One of the most infamous chemical warfare agents of all time, mustard gas (dichlorodiethyl sulfide), is also not listed, and is (at least theoretically) not regulated as hazardous under RCRA! Because there are some 6 million chemical compounds identified, it would be almost impossible to regulate every chemical that shows some sign of being hazardous. Nonetheless, a greater effort to provide a good definition of toxic must be made. Toxicologists have struggled to more clearly define what is toxic and what is not and have begun to work with government to develop specific risk assessments that interpret this data.

There is little indication that any significant number of wastes will be delisted in the future. Each effort by a specific industry to deregulate a waste material has been met by opposition. Economic concerns have generally fallen on deaf ears. Industry must be prepared to either change processes, minimize the quantity of wastes generated, or pay the price for proper disposal of those wastes that are now listed or meet defined characteristics.

WASTE MINIMIZATION

A number of studies into waste minimization practices have been published, including a few government studies. It might well be practical to write an entire book on the subject. Referred to almost interchangeably as waste reduction, waste minimization is the most obvious way for all hazardous waste generators to reduce costs for disposal.

Small quantity generators can certainly reduce their costs by reducing the quantities of waste generated. The most logical ways to reduce quantities are to change processes to use less hazardous materials and to improve the efficiency of processes that generate waste. Proper purchasing and raw material inventory control procedures are also important in a waste minimization program. Currently, large waste generators are required to sign a statement on the hazardous waste manifest that says they have a waste minimization program in place and are doing everything possible to reduce the quantities of waste they generate. While there is currently no federal requirement for such a certification by small quantity generators, it is likely that there will be a continued focus at both the federal and state levels on efforts to reduce quantities of waste.

GREATER FOCUS ON NONHAZARDOUS WASTES (SOLID WASTE)

With landfill space disappearing at a rapid pace for the disposal of all types of solid waste, alternate technologies for the disposal of nonhazardous wastes are also being researched. The best solution, in many cases, appears to be the use of trash-to-steam plants. Baltimore is among the first large cities to build and operate such a facility, and all indications are that the project is a success. Other large cities in the Northeast, in particular, will soon be forced to consider such strategies. New York and Philadelphia both have plans on the drawing board to develop trash-to-energy facilities. With ash from incinerators still representing a disposal problem in most cases, valuable landfill space might eventually be reserved solely for the residues of other disposal technologies.

NIMBY SYNDROME: A CHALLENGE FOR SOCIETY

Few communities will support the idea of hazardous waste facilities being sited in their region. The perceived notion of increased health problems, the noise and potential hazard associated with transportation of hazardous waste, and the assumed decrease in property values and quality of life are all deterrents to siting of facilities. Nonetheless, the "Not In My Back Yard" syndrome must be overcome for hazardous waste to be managed properly. As more and more facilities are forced to close, those remaining will strain to handle additional materials. Transportation requirements will increase, leaving greater potential for environmental incidents.

Most states have developed siting standards for disposal facilities, but few have overcome political pressures to the extent that specific sites have been selected for future sites. It is largely up to industry to campaign for properly managed regional sites, even if political pressures discourage siting. This is a problem we all face, and small businesses in particular cannot afford to handle the burden of hazardous waste disposal costs increasing at their current pace. Businesses and jobs may be lost in areas that do not have access to good waste management facilities.

APPENDIX I

Hazardous Waste Agencies and Small Quantity Limits

EPA REGIONAL OFFICES

EPA Region 1
State Waste Programs Branch
JFK Federal Building
Boston, MA 02203
(617) 223-3468
Connecticut, Massachusetts, Maine,
New Hampshire, Rhode Island,
Vermont

EPA Region 2
Air and Waste Management
Division
26 Federal Plaza
New York, NY 10278
(212) 264-5175
New Jersey, New York, Puerto
Rico, Virgin Islands

EPA Region 3
Waste Management Branch
841 Chestnut Street
Philadelphia, PA 19107
(215) 597-9336
Delaware, Maryland, Pennsylvania,
Virgina, West Virginia, District
of Columbia

EPA Region 4
Hazardous Waste Management
Division
345 Courtland Street, N.E.
Atlanta, GA 30365
(404) 347-3016
Alabama, Florida, Georgia, Kentucky, Mississippi, North Carolina, South Carolina, Tennessee

EPA Region 5
RCRA Activities
230 S. Dearborn Street
Chicago, IL 60604
(312) 353-2000
Illinois, Indiana, Michigan,
    Minnesota, Ohio, Wisconsin

EPA Region 6
Air and Hazardous Materials
    Division
1201 Elm Street
Dallas, TX 75270
(214) 767-2600
Arkansas, Louisiana, New Mexico,
    Oklahoma, Texas

EPA Region 7
RCRA Branch
726 Minnesota Avenue
Kansas City, KS 66101
(913) 236-2800
Iowa, Kansas, Missouri, Nebraska

EPA Region 8
Waste Management Division
    (8HWM-ON)
One Denver Place

999 18th Street, Suite 1300
Denver, CO 80202–2413
(303) 293-1502
Colorado, Montana, North
    Dakota, South Dakota, Utah,
    Wyoming

EPA Region 9
Toxics and Waste Management
    Division
215 Fremont Street
San Francisco, CA 94105
(415) 974-7472
Arizona, California, Hawaii,
    Nevada, American Samoa,
    Guam, Trust Territories of the
    Pacific

EPA Region 10
Waste Management Branch —
    MS-530
1200 Sixth Avenue
Seattle, WA 98101
(206) 442-2777
Alaska, Idaho, Oregon,
    Washington

STATE AGENCIES*

Alabama
Alabama Department of Environ-
    mental Management
Land Division
1751 Federal Drive
Montgomery, AL 36130
(205) 271-7730
SMALL QUANTITY LIMITS: SAME AS
    FEDERAL

Alaska
Department of Environmental Con-
    servation
P.O. Box O
Juneau, AK 99811
(907) 465-2666
SMALL QUANTITY LIMITS: SAME AS
    FEDERAL

---

*States which currently have fully authorized programs are marked with an asterisk.

Arizona*
Arizona Department of Health
  Services
Office of Waste and Water Quality
2005 N. Central Avenue, Room 304
Phoenix, AZ 85004
(602) 255-2211
SMALL QUANTITY LIMITS: SAME AS
  FEDERAL

Arkansas*
Department of Pollution Control
  and Ecology
Hazardous Waste Division
P.O. Box 9583
8001 National Drive
Little Rock, AR 72219
(501) 562-7444
SMALL QUANTITY LIMITS: SAME AS
  FEDERAL

California*
Department of Health Services
Toxic Substances Control Division
714 P Street, Room 1253
Sacramento, CA 95814
(916) 324-1826
SMALL QUANTITY LIMITS: NO
  EXEMPTIONS FOR SMALL QUAN-
  TITY GENERATORS

Colorado*
Colorado Department of Health
Waste Management Division
4210 E. 11th Avenue
Denver, CO 80220
(303) 320-8333 ext. 4364
SMALL QUANTITY LIMITS: SAME AS
  FEDERAL

Connecticut
Department of Environmental Protection

Hazardous Waste Management
  Section
State Office Building
165 Capitol Avenue
Hartford, CT 06106
(203) 566-8843
SMALL QUANTITY LIMITS: 1000 KG
  (NO ACUTE OR SPILL RESIDUE
  PROVISIONS)

Delaware*
Department of Natural Resources
  and Environmental Control
Waste Management Section
P.O. Box 1401
Dover, DE 19903
(302) 736-4781
SMALL QUANTITY LIMITS: SAME AS
  FEDERAL

District of Columbia
Department of Consumer and Reg-
  ulatory Affairs
Pesticides and Hazardous Waste
  Materials Division
5010 Overlook Avenue, S.W.,
  Room 114
Washington, DC 20032
(202) 767-8414
SMALL QUANTITY LIMITS: SAME AS
  FEDERAL

Florida*
Department of Environmental Reg-
  ulation
Solid and Hazardous Waste Section
Twin Towers Office Building
2600 Blair Stone Road
Tallahassee, FL 32301
(904) 488-0300
SMALL QUANTITY LIMITS: SAME AS
  FEDERAL

Georgia*
Georgia Environmental Protection
  Division
Hazardous Waste Management
  Program
Land Protection Branch
Floyd Towers East, Suite 1154
205 Butler Street, S.E.
Atlanta, GA 30334
(404) 656-2833 or (800) 334-2373
SMALL QUANTITY LIMITS: SAME AS
  FEDERAL

Hawaii
Department of Health
Environmental Health Division
P.O. Box 3378
Honolulu, HI 96801
(808) 548-4383
SMALL QUANTITY LIMITS: SAME AS
  FEDERAL

Idaho
Department of Health and Welfare
Bureau of Hazardous Materials
450 W. State Street
Boise, ID 83720
(208) 334-5879
SMALL QUANTITY LIMITS: SAME AS
  FEDERAL

Illinois*
Environmental Protection Agency
Division of Land Pollution Control
2200 Churchill Road, #24
Springfield, IL 62706
(217) 782-6761
SMALL QUANTITY LIMITS: 100 KG
  FOR NONACUTE WASTE, 1 KG
  FOR HIGHLY ACUTE WASTE

Indiana*
Department of Environmental
  Management

Office of Solid and Hazardous
  Waste
105 S. Meridian
Indianapolis, IN 46225
(317) 232-4535
SMALL QUANTITY LIMITS: SAME AS
  FEDERAL

Iowa
U.S. EPA Region 7
Hazardous Materials Branch
726 Minnesota Avenue
Kansas City, KS 66101
(913) 236-2888 Iowa RCRA Toll
  Free: (800) 223-0425
SMALL QUANTITY LIMITS: SAME AS
  FEDERAL

Kansas*
Department of Health and Envi-
  ronment
Bureau of Waste Management
Forbes Field, Bldg. 321
Topeka, KS 66620
(913) 862-9360 ext. 292
SMALL QUANTITY LIMITS: 25 KG/
  MONTH FOR NONACUTE WASTE
  AND SPILL RESIDUES, 1 KG/
  MONTH FOR ACUTE WASTE

Kentucky*
Natural Resources and Environ-
  mental Protection Cabinet
Division of Waste Management
18 Reilly Road
Frankfort, KY 40601
(502) 564-6716
SMALL QUANTITY LIMITS: SAME AS
  FEDERAL

Louisiana*
Department of Environmental
  Quality
Hazardous Waste Division

P.O. Box 44307
Baton Rouge, LA 70804
(504) 342-1227
SMALL QUANTITY LIMITS: GENER-
ATE UP TO 100 KG/MONTH,
ACCUMULATE UP TO 1000 KG.
ONCE AT THAT LIMIT, DISPOSE
OF IN 90 DAYS. NO STORAGE
BEYOND ONE YEAR UNDER ANY
CIRCUMSTANCES.

Maine

Department of Environmental Pro-
tection
Bureau of Oil and Hazardous
Materials Control
State House Station #17
Augusta, ME 04333
(207) 289-2951
SMALL QUANTITY LIMITS: NON-
ACUTE WASTES 100 KG, ACUTE
WASTES 1 KG, SAME LIMITS FOR
SPILL MATERIAL

Maryland*

Department of Health and Mental
Hygiene
Maryland Waste Management
Administration
Office of Environmental Programs
201 W. Preston Street, Room A3
Baltimore, MD 21201
(301) 225-5709
SMALL QUANTITY LIMITS: 1 KG
ACUTE WASTE, ALL ACUTE
CONTAINERS OVER 20 LITERS,
10 KG OF INNER LINERS CON-
TAMINATED WITH ACUTE
WASTE, 100 KG OF CONTAMI-
NATED SOIL, 1000 KG OF NON-
ACUTE WASTE

Massachusetts*

Department of Environmental
Quality Engineering
Division of Solid and Hazardous
Waste
One Winter Street, 5th floor
Boston, MA 02108
(617) 292-5589, (617) 292-5851
SMALL QUANTITY LIMITS: 1 KG
ACUTE WASTE, 100 KG NON-
ACUTE WASTE

Michigan

Michigan Department of Natural
Resources
Hazardous Waste Division
Waste Evaluation Unit
Box 30028
Lansing, MI 48909
(517) 373-2730
SMALL QUANTITY LIMITS: SAME AS
FEDERAL

Minnesota*

Pollution Control Agency
Solid and Hazardous Waste Divi-
sion
1935 W. County Road, B-2
Roseville, MN 55113
(612) 296-7282
SMALL QUANTITY LIMITS: NO
EXEMPTIONS FOR SMALL QUAN-
TITY GENERATORS

Mississippi*

Department of Natural Resources
Division of Solid and Hazardous
Waste Management
P.O. Box 10385
Jackson, MS 39209
(601) 961-5062
SMALL QUANTITY LIMITS: SAME AS
FEDERAL

Missouri*

Department of Natural Resources
Waste Management Program
P.O. Box 176
Jefferson City, MO 65102
(314) 751-3176 Hotline: (800) 334-6946
SMALL QUANTITY LIMITS: SAME AS FEDERAL

Montana*

Department of Health and Environmental Sciences
Solid and Hazardous Waste Bureau
Cogswell Building, Room B-201
Helena, MT 59620
(406) 444-2821
SMALL QUANTITY LIMITS: SAME AS FEDERAL

Nebraska*

Department of Environmental Control
Hazardous Waste Management Section
P.O. Box 94877
State House Station
Lincoln, NE 68509
(402) 471-2186
SMALL QUANTITY LIMITS: SAME AS FEDERAL

Nevada*

Division of Environmental Protection
Waste Management Program
Capitol Complex
Carson City, NV 89710
(702) 885-4670
SMALL QUANTITY LIMITS: SAME AS FEDERAL

New Hampshire*

Department of Health and Human Services
Office of Waste Management
Health and Welfare Building
Hazen Drive
Concord, NH 03301–6527
(603) 271-4608
SMALL QUANTITY LIMITS: 1 KG ACUTE, 100 KG SPILL RESIDUE, 100 KG NONACUTE WASTE

New Jersey*

Department of Environmental Protection
Division of Waste Management
32 E. Hanover Street, CN-028
Trenton, NJ 08625
(609) 292-8341
SMALL QUANTITY LIMITS: 1 KG ACUTE, 100 KG SPILL RESIDUE, 100 KG NONACUTE WASTE

New Mexico*

Environmental Improvement Division
Ground Water and Hazardous Waste Bureau
Hazardous Waste Section
P.O. Box 968
Santa Fe, NM 87504–0968
(505) 827-2922
SMALL QUANTITY LIMITS: SAME AS FEDERAL

New York*

Department of Environmental Conservation
Bureau of Hazardous Waste Operations
50 Wolf Road, Room 209
Albany, NY 12233
(518) 457-0530 SQG HOTLINE: (800) 631-0666

SMALL QUANTITY LIMITS: SAME AS
FEDERAL

North Carolina*
Department of Human Resources
Solid and Hazardous Waste Management Branch
P.O. Box 2091
Raleigh, NC 27602
(919) 733-2178
SMALL QUANTITY LIMITS: SAME AS
FEDERAL. CONDITIONALLY
EXEMPT GENERATORS CANNOT
SEND HAZARDOUS WASTE TO
MUNICIPAL LANDFILLS.

North Dakota*
Department of Health
Division of Hazardous Waste Management and Special Studies
1200 Missouri Avenue
Bismarck, ND 58502-5520
(701) 224-2366
SMALL QUANTITY LIMITS: SAME AS
FEDERAL

Ohio
Ohio EPA
Division of Solid and Hazardous Waste Management
361 E. Broad Street
Columbus, OH 43266-0558
(614) 466-7220
SMALL QUANTITY LIMITS: SAME AS
FEDERAL

Oklahoma*
Waste Management Service
Oklahoma State Department of Health
P.O. Box 53551
Oklahoma City, OK 73152
(405) 271-5338

SMALL QUANTITY LIMITS: SAME AS
FEDERAL

Oregon*
Hazardous and Solid Waste Division
P.O. Box 1760
Portland, OR 97207
(503) 229-6534
SMALL QUANTITY LIMITS: SAME AS
FEDERAL

Pennsylvania*
Department of Environmental Resources
Bureau of Waste Management
P.O. Box 2063
Harrisburg, PA 17120
(717) 787-6239
SMALL QUANTITY LIMITS: SAME AS
FEDERAL

Puerto Rico
Environmental Quality Board
P.O. Box 11488
Santurce, PR 00910
(809) 723-8184
SMALL QUANTITY LIMITS: SAME AS
FEDERAL

Rhode Island*
Department of Environmental Management
Division of Air and Hazardous Materials
Room 204, Cannon Building
75 Davis Street
Providence, RI 02908
SMALL QUANTITY LIMITS: NO
EXEMPTIONS FOR SMALL QUANTITY GENERATORS

South Carolina*

Department of Health and Environmental Control
Bureau of Solid and Hazardous Waste Management
2600 Bull Street
Columbia, SC 29201
(803) 734-5200
SMALL QUANTITY LIMITS: LESS THAN 100 KG EXEMPT, OVER 100 KG REGULATED AS LARGE GENERATOR

South Dakota*

Department of Water and Natural Resources
Office of Air Quality and Solid Waste
Foss Building, Room 217
Pierre, SD 57501
(605) 773-3153
SMALL QUANTITY LIMITS: SAME AS FEDERAL

Tennessee*

Division of Solid Waste Management
Tennessee Department of Public Health
701 Broadway
Nashville, TN 37219–5403
(615) 741-3424
SMALL QUANTITY LIMITS: SAME AS FEDERAL

Texas*

Texas Water Commission
Hazardous and Solid Waste Division
Program Support Section
1700 N. Congress
Austin, TX 78711
(512) 463-7761
SMALL QUANTITY LIMITS: SAME AS

FEDERAL (PROPOSED RULE: NO CONDITIONAL EXEMPTIONS)

Utah*

Department of Health
Bureau of Solid and Hazardous Waste Management
P.O. Box 16700
Salt Lake City, UT 84116–0700
(801) 538-6170
SMALL QUANTITY LIMITS: SAME AS FEDERAL

Vermont*

Agency of Environmental Conservation
103 S. Main Street
Waterbury, VT 05676
(802) 244-8702
SMALL QUANTITY LIMITS: 1 KG ACUTE WASTE, 100 KG NON-ACUTE WASTE; SAME FOR SPILL RESIDUES

Virginia*

Department of Health
Division of Solid and Hazardous Waste Management
Monroe Building, 11th floor
101 N. 14th Street
Richmond, VA 23219
(804) 225-2667 Hotline: (800) 552-2075
SMALL QUANTITY LIMITS: SAME AS FEDERAL

Washington*

Department of Ecology
Solid and Hazardous Waste Program
Mail Stop PV-11
Olympia, WA 98504–8711
(206) 459-6322 or (800) 633-7585 in state

SMALL QUANTITY LIMITS: SAME AS
  FEDERAL

West Virginia
Division of Water Resources
Solid and Hazardous Waste/
  Ground Water Branch
1201 Greenbrier Street
Charleston, WV 25311
SMALL QUANTITY LIMITS: SAME AS
  FEDERAL

Wisconsin*
Department of Natural Resources
Bureau of Solid Waste Manage-
  ment

P.O. Box 7921
Madison, WI 53707
(608) 266-1327
SMALL QUANTITY LIMITS: SAME AS
  FEDERAL

Wyoming
Department of Environmental
  Quality
Solid Waste Management Program
122 W. 25th Street
Cheyenne, WY 82002
(307) 777-7752
SMALL QUANTITY LIMITS: SAME AS
  FEDERAL

APPENDIX II

Materials Which Are Not Solid Wastes

Source: Reprinted from 40 CFR 261.4 (a)(1–7).

261.4 EXCLUSIONS.

(a) Materials which are not solid wastes. The following materials are not solid wastes for the purpose of this part:

(1)(i) Domestic sewage; and

(ii) Any mixture of domestic sewage and other wastes that passes through a sewer system to a publicly-owned treatment works for treatment. "Domestic Sewage" means untreated sanitary wastes that pass through a sewer system.

(2) Industrial wastewater discharges that are point source discharges subject to regulation under Section 402 of the Clean Water Act, as amended.

[Comment: This exclusion applies only to the actual point source discharge. It does not exclude industrial wastewaters while they are being collected, stored or treated before discharge, nor does it exclude sludges that are generated by industrial wastewater treatment.]

(3) Irrigation return flows.

(4) Source, special nuclear or byproduct material as defined by the Atomic Energy Act of 1954, as amended, 42 U.S.C.2011 et seq.

(5) Materials subjected to in-situ mining techniques which are not removed from the ground as part of the extraction process.

(6) Pulping liquors (ie., black liquor) that are reclaimed in the pulping liquor recovery furnace and then reused in the pulping process, unless it is accumulated speculatively as defined in 261.1(c) of this chapter.

(7) Spent sulfuric acid used to produce virgin sulfuric acid, unless it is accumulated speculatively as defined in 261.1(c) of this chapter.

APPENDIX III

Solid Wastes Which Are Not Hazardous Wastes

Source: Reprinted from 40 CFR 261.4 (b), (c), and (d)

(b) *Solid wastes which are not hazardous wastes.* The following solid wastes are not hazardous wastes:

(1) Household waste, including household waste that has been collected, transported, stored, treated, disposed, recovered (*e.g.*, refuse-derived fuel), or reused. "Household waste" means any waste material (including garbage, trash and sanitary wastes in septic tanks) derived from households (including single and multiple residences, hotels and motels, bunkhouses, ranger stations, crew quarters, campgrounds, picnic grounds, and day-use recreation areas).

(2) Solid wastes generated by any of the following and which are returned to the soils as fertilizers:

(i) The growing and harvesting of agricultural crops.

(ii) The raising of animals, including animal manures.

(3) Mining overburden returned to the mine site.

(4) Fly ash waste, bottom ash waste, slag waste, and flue gas emission control waste generated primarily from the combustion of coal or other fossil fuels.

(5) Drilling fluids, produced waters, and other wastes associated

with the exploration, development, or production of crude oil, natural gas or geothermal energy.

(6)(i) Wastes which fail the test for the characteristic of EP toxicity because chromium is present or are listed in Subpart D due to the presence of chromium, which do not fail the test for the characteristic of EP toxicity for any other constituent or are not listed due to the presence of any other constituent, and which do not fail the test for any other characteristic, if it is shown by a waste generator or by waste generators that:

(A) The chromium in the waste is exclusively (or nearly exclusively) trivalent chromium; and

(B) The waste is generated from an industrial process which uses trivalent chromium exclusively (or nearly exclusively) and the process does not generate hexavalent chromium; and

(C) The waste is typically and frequently managed in non-oxidizing environments.

(ii) Specific wastes which meet the standard in paragraphs (b)(6)(i)(A), (B) and (C) (so long as they do not fail the test for the characteristic of EP toxicity, and do not fail the test for any other characteristic) are:

(A) Chrome (blue) trimmings generated by the following subcategories of the leather tanning and finishing industry; hair pulp/chrome tan/retan/wet finish; hair save/chrome tan/retan/wet finish; retan/wet finish; no beamhouse; through-the-blue; and shearling.

(B) Chrome (blue) shavings generated by the following subcategories of the leather tanning and finishing industry; hair pulp/chrome tan/retan/wet finish; hair save/chrome tan/retan/wet finish; retan/wet finish; no beamhouse; through-the-blue; and shearling.

(C) Buffing dust generated by the following subcategories of the leather tanning and finishing industry; hair pulp/chrome tan/retan/wet finish; hair save/chrome tan/retan/wet finish; retan/wet finish; no beamhouse; through-the-blue.

(D) Sewer screenings generated by the following subcategories of the leather tanning and finishing industry: hair pulp/chrome tan/retan/wet finish; hair save/chrome tan/retan/wet finish; retan/wet finish; no beamhouse; through-the-blue; and shearling.

(E) Wastewater treatment sludges generated by the following subcategories of the leather tanning and finishing industry; hair pulp/chrome tan/retan/wet finish; hair save/chrome tan/retan/wet finish; retan/wet finish; no beamhouse; through-the-blue; and shearling.

(F) Wastewater treatment sludges generated by the following subcategories of the leather tanning and finishing industry: hair pulp/ chrome tan/retan/wet finish; hair save/chrome-tan/retan/wet finish; and through-the-blue.

(G) Waste scrap leather from the leather tanning industry, the shoe manufacturing industry, and other leather product manufacturing industries.

(H) Wastewater treatment sludges from the production of TiO_2, pigment using chromium-bearing ores by the chloride process.

(7) Solid waste from the extraction, beneficiation and processing of ores and minerals (including coal), including phosphate rock and overburden from the mining of uranium ore.

(8) Cement kiln dust waste.

(9) Solid waste which consists of discarded wood or wood products which fails the test for the characteristic of EP toxicity and which is not a hazardous waste for any other reason if the waste is generated by persons who utilize the arsenical-treated wood and wood products for these materials' intended end use.

(c) Hazardous wastes which are exempted from certain regulations. A hazardous waste which is generated in a product or raw material storage tank, a product or raw material transport vehicle or vessel, a product or raw material pipeline, or in a manufacturing process unit or an associated nonwaste-treatment-manufacturing unit, is not subject to regulation under Parts 262 through 265, 270, 271 and 124 of this chapter or to the notification requirements of Section 3010 of RCRA until it exits the unit in which it was generated, unless the hazardous waste remains in the unit more than 90 days after the unit ceases to be operated for manufacturing, or for storage or transportation of product or raw materials.

(d) *Samples.* (1) Except as provided in paragraph (d)(2) of this section, a sample of solid waste or a sample of water, soil, or air, which is collected for the sole purpose of testing to determine its characteristics or composition, is not subject to any requirements of this part or Parts 262 through 267 or Part 270 or Part 124 of this chapter or to the notification requirements of Section 3010 of RCRA, when:

(i) The sample is being transported to a laboratory for the purpose of testing; or

(ii) The sample is being transported back to the sample collector after testing; or

(iii) The sample is being stored by the sample collector before transport to a laboratory for testing; or

(iv) The sample is being stored in a laboratory before testing; or

(v) The sample is being stored in a laboratory after testing but before it is returned to the sample collector; or

(vi) The sample is being stored temporarily in the laboratory after testing for a specific purpose (for example, until conclusion of a court case or enforcement action where further testing of the sample may be necessary).

(2) In order to qualify for the exemption in paragraphs (d)(1) (i) and (ii) of this section, a sample collector shipping samples to a laboratory and a laboratory returning samples to a sample collector must:

(i) Comply with U.S. Department of Transportation (DOT), U.S. Postal Service (USPS), or any other applicable shipping requirements; or

(ii) Comply with the following requirements if the sample collector determines that DOT, USPS, or other shipping requirements do not apply to the shipment of the sample:

(A) Assure that the following information accompanies the sample:

(*1*) The sample collector's name, mailing address, and telephone number;

(*2*) The laboratory's name, mailing address, and telephone number;

(*3*) The quantity of the sample;

(*4*) The date of shipment; and

(*5*) A description of the sample.

(B) Package the sample so that it does not leak, spill, or vaporize from its packaging.

(3) This exemption does not apply if the laboratory determines that the waste is hazardous but the laboratory is no longer meeting any of the conditions stated in paragraph (d)(1) of this section.

APPENDIX IV

Listed Hazardous Wastes

Source: Reprinted from 40 CFR 261.31, 261.32, and 261.33.

SUBPART D—LISTS OF HAZARDOUS WASTES

§ 261.30 General.

(a) A solid waste is a hazardous waste if it is listed in this subpart, unless it has been excluded from this list under §§ 260.20 and 260.22.

(b) The Administrator will indicate his basis for listing the classes or types of wastes listed in this Subpart by employing one or more of the following Hazard Codes:

Ignitable Waste .... (I)
Corrosive Waste .... (C)
Reactive Waste .... (R)
EP Toxic Waste .... (E)
Acute Hazardous Waste .... (H)
Toxic Waste .... (T)

Appendix VII identifies the constituent which caused the Administrator to list the waste as an EP Toxic Waste (E) or Toxic Waste (T) in §§ 261.31 and 261.32.

(c) Each hazardous waste listed in this subpart is assigned an EPA

Hazardous Waste Number which precedes the name of the waste. This number must be used in complying with the notification requirements of Section 3010 of the Act and certain recordkeeping and reporting requirements under Parts 262 through 265 and Part 270 of this chapter.

(d) The following hazardous wastes listed in § 261.31 or § 261.32 are subject to the exclusion limits for acutely hazardous wastes established in § 261.5: EPA Hazardous Wastes Nos. FO20, FO21, FO22, FO23, FO26, and FO27.

[45 FR 33119, May 19, 1980, as amended at 48 FR 14294, Apr. 1, 1983; 50 FR 2000, Jan. 14, 1985]

§261.31 Hazardous wastes from non-specific sources.

The following solid wastes are listed hazardous wastes from non-specific sources unless they are excluded under §§ 260.20 and 260.22 and listed in Appendix IX.

Industry and EPA hazardous waste No.	Hazardous waste	Hazard code
Generic: F001	The following spent halogenated solvents used in degreasing: Tetrachloroethylene, trichloroethylene, methylene chloride, 1,1,1-trichloroethane, carbon tetrachloride, and chlorinated fluorocarbons; all spent solvent mixtures/blends used in degreasing containing, before use, a total of ten per cent or more (by volume) of one or more of the above halogenated solvents or those solvents listed in F002, F004, and F005; and still bottoms from the recovery of these spent solvents and spent solvent mixtures.	(T)
F002	The following spent halogenated solvents: Tetrachloroethylene, methylene chloride, trichloroethylene, 1,1,1-trichloroethane, chlorobenzene, 1,1,2-trichloro-1,2,2-trifluoroethane, ortho-dichlorobenzene, trichlorofluoromethane, and 1,1,2-trichloroethane; all spent solvent mixtures/blends containing, before use, a total of ten percent or more (by volume) of one or more of the above halogenated solvents or those listed in F001, F004, or F005; and still bottoms from the recovery of these spent solvents and spent solvent mixtures.	(T)
F003	The following spent non-halogenated solvents: Xylene, acetone, ethyl acetate, ethyl benzene, ethyl ether, methyl isobutyl ketone, n-butyl alcohol, cyclohexanone, and methanol; all spent solvent mixtures/blends containing, before use, only the above spent non-halogenated solvents; and all spent solvent mixtures/blends containing, before use, one or more of the above non-halogenated solvents, and, a total of ten percent or more (by volume) of one or more of those solvents listed in F001, F002, F004, and F005; and still bottoms from the recovery of these spent solvents and spent solvent mixtures.	(I)*
F004	The following spent non-halogenated solvents: Cresols and cresylic acid, and nitrobenzene; all spent solvent mixtures/blends containing, before use, a total of ten percent or more (by volume) of one or more of the above non-halogenated solvents or those solvents listed in F001, F002, and F005; and still bottoms from the recovery of these spent solvents and spent solvent mixtures.	(T)
F005	The following spent non-halogenated solvents: Toluene, methyl ethyl ketone, carbon disulfide, isobutanol, pyridine, benzene, 2-ethoxyethanol, and 2-nitropropane; all spent solvent mixtures/blends containing, before use, a total of ten percent or more (by volume) of one or more of the above non-halogenated solvents or those solvents listed in F001, F002, or F004; and still bottoms from the recovery of these spent solvents and spent solvent mixtures.	(I,T)
F006	Wastewater treatment sludges from electroplating operations except from the following processes: (1) Sulfuric acid anodizing of aluminum; (2) tin plating on carbon steel; (3) zinc plating (segregated basis) on carbon steel; (4) aluminum or zinc-aluminum plating on carbon steel; (5) cleaning/stripping associated with tin, zinc and aluminum plating on carbon steel; and (6) chemical etching and milling of aluminum.	(T)

Industry and EPA hazardous waste No.	Hazardous waste	Hazard code
F019	Wastewater treatment sludges from the chemical conversion coating of aluminum	(T)
F007	Spent cyanide plating bath solutions from electroplating operations	(R, T)
F008	Plating bath residues from the bottom of plating baths from electroplating operations where cyanides are used in the process.	(R, T)
F009	Spent stripping and cleaning bath solutions from electroplating operations where cyanides are used in the process.	(R, T)
F010	Quenching bath residues from oil baths from metal heat treating operations where cyanides are used in the process.	(R, T)
F011	Spent cyanide solutions from salt bath pot cleaning from metal heat treating operations.	(R, T)
F012	Quenching waste water treatment sludges from metal heat treating operations where cyanides are used in the process.	(T)
F024	Wastes, including but not limited to, distillation residues, heavy ends, tars, and reactor clean-out wastes from the production of chlorinated aliphatic hydrocarbons, having carbon content from one to five, utilizing free radical catalyzed processes. (This listing does not include light ends, spent filters and filter aids, spent dessicants, wastewater, wastewater treatment sludges, spent catalysts, and wastes listed in § 261.32.].	(T)
F020	Wastes (except wastewater and spent carbon from hydrogen chloride purification) from the production or manufacturing use (as a reactant, chemical intermediate, or component in a formulating process) of tri- or tetrachlorophenol, or of intermediates used to produce their pesticide derivatives. (This listing does not include wastes from the production of Hexachlorophene from highly purified 2,4,5-trichlorophenol.).	(H)
F021	Wastes (except wastewater and spent carbon from hydrogen chloride purification) from the production or manufacturing use (as a reactant, chemical intermediate, or component in a formulating process) of pentachlorophenol, or of intermediates used to produce its derivatives.	(H)
F022	Wastes (except wastewater and spent carbon from hydrogen chloride purification) from the manufacturing use (as a reactant, chemical intermediate, or component in a formulating process) of tetra-, penta-, or hexachlorobenzenes under alkaline conditions.	(H)
F023	Wastes (except wastewater and spent carbon from hydrogen chloride purification) from the production of materials on equipment previously used for the production or manufacturing use (as a reactant, chemical intermediate, or component in a formulating process) of tri- and tetrachlorophenols. (This listing does not include wastes from equipment used only for the production or use of Hexachlorophene from highly purified 2,4,5-trichlorophenol.).	(H)
F026	Wastes (except wastewater and spent carbon from hydrogen chloride purification) from the production of materials on equipment previously used for the manufacturing use (as a reactant, chemical intermediate, or component in a formulating process) of tetra-, penta-, or hexachlorobenzene under alkaline conditions.	(H)
F027	Discarded unused formulations containing tri-, tetra-, or pentachlorophenol or discarded unused formulations containing compounds derived from these chlorophenols. (This listing does not include formulations containing Hexachlorophene sythesized from prepurified 2,4,5-trichlorophenol as the sole component.).	(H)
F028	Residues resulting from the incineration or thermal treatment of soil contaminated with EPA Hazardous Waste Nos. F020, F021, F022, F023, F026, and F027.	(T)

*(I,T) should be used to specify mixtures containing ignitable and toxic constituents.

[46 FR 4617, Jan. 16, 1981, as amended at 46 FR 27477, May 20, 1981; 49 FR 5312, Feb. 10, 1984; 49 FR 37070, Sept. 21, 1984; 50 FR 665, Jan. 4, 1985; 50 FR 2000, Jan. 14, 1985; 50 FR 53319, Dec. 31, 1985; 51 FR 2702, Jan. 21, 1986; 51 FR 6541, Feb. 25, 1986]

EFFECTIVE DATE NOTE: At 51 FR 6541, Feb. 25, 1986, in § 261.31, waste streams "F002" and "F005" in the subgroup "Generic" were revised, effective August 25, 1986. For the convenience of the user, the superseded text is set forth as follows:

§ 261.32 Hazardous wastes from specific sources.

The following solid wastes are listed hazardous wastes from specific sources unless they are excluded under §§ 260.20 and 260.22 and listed in Appendix IX.

Industry and EPA hazardous waste No.	Hazardous waste	Hazard code
Wood preservation: K001	Bottom sediment sludge from the treatment of wastewaters from wood preserving processes that use creosote and/or pentachlorophenol.	(T)
Inorganic pigments:		
K002	Wastewater treatment sludge from the production of chrome yellow and orange pigments.	(T)
K003	Wastewater treatment sludge from the production of molybdate orange pigments	(T)
K004	Wastewater treatment sludge from the production of zinc yellow pigments	(T)
K005	Wastewater treatment sludge from the production of chrome green pigments	(T)
K006	Wastewater treatment sludge from the production of chrome oxide green pigments (anhydrous and hydrated).	(T)

Industry and EPA hazardous waste No.	Hazardous waste	Hazard code
K007	Wastewater treatment sludge from the production of iron blue pigments	(T)
K008	Oven residue from the production of chrome oxide green pigments	(T)
Organic chemicals:		
K009	Distillation bottoms from the production of acetaldehyde from ethylene	(T)
K010	Distillation side cuts from the production of acetaldehyde from ethylene	(T)
K011	Bottom stream from the wastewater stripper in the production of acrylonitrile	(R, T)
K013	Bottom stream from the acetonitrile column in the production of acrylonitrile	(R, T)
K014	Bottoms from the acetonitrile purification column in the production of acrylonitrile	(T)
K015	Still bottoms from the distillation of benzyl chloride	(T)
K016	Heavy ends or distillation residues from the production of carbon tetrachloride	(T)
K017	Heavy ends (still bottoms) from the purification column in the production of epichlorohydrin.	(T)
K018	Heavy ends from the fractionation column in ethyl chloride production	(T)
K019	Heavy ends from the distillation of ethylene dichloride in ethylene dichloride production.	(T)
K020	Heavy ends from the distillation of vinyl chloride in vinyl chloride monomer production.	(T)
K021	Aqueous spent antimony catalyst waste from fluoromethanes production	(T)
K022	Distillation bottom tars from the production of phenol/acetone from cumene	(T)
K023	Distillation light ends from the production of phthalic anhydride from naphthalene	(T)
K024	Distillation bottoms from the production of phthalic anhydride from naphthalene	(T)
K093	Distillation light ends from the production of phthalic anhydride from ortho-xylene	(T)
K094	Distillation bottoms from the production of phthalic anhydride from ortho-xylene	(T)
K025	Distillation bottoms from the production of nitrobenzene by the nitration of benzene	(T)
K026	Stripping still tails from the production of methy ethyl pyridines	(T)
K027	Centrifuge and distillation residues from toluene diisocyanate production	(R, T)
K028	Spent catalyst from the hydrochlorinator reactor in the production of 1,1,1-trichloroethane.	(T)
K029	Waste from the product steam stripper in the production of 1,1,1-trichloroethane	(T)
K095	Distillation bottoms from the production of 1,1,1-trichloroethane	(T)
K096	Heavy ends from the heavy ends column from the production of 1,1,1-trichloroethane.	(T)
K030	Column bottoms or heavy ends from the combined production of trichloroethylene and perchloroethylene.	(T)
K083	Distillation bottoms from aniline production	(T)
K103	Process residues from aniline extraction from the production of aniline	(T)
K104	Combined wastewater streams generated from nitrobenzene/aniline production	(T)
K085	Distillation or fractionation column bottoms from the production of chlorobenzenes	(T)
K105	Separated aqueous stream from the reactor product washing step in the production of chlorobenzenes.	(T)
K111	Product washwaters from the production of dinitrotoluene via nitration of toluene	(C,T)
K112	Reaction by-product water from the drying column in the production of toluenediamine via hydrogenation of dinitrotoluene.	(T)
K113	Condensed liquid light ends from the purification of toluenediamine in the production of toluenediamine via hydrogenation of dinitrotoluene.	(T)
K114	Vicinals from the purification of toluenediamine in the production of toluenediamine via hydrogenation of dinitrotoluene.	(T)
K115	Heavy ends from the purification of toluenediamine in the production of toluenediamine via hydrogenation of dinitrotoluene.	(T)
K116	Organic condensate from the solvent recovery column in the production of toluene diisocyanate via phosgenation of toluenediamine.	(T)
K117	Wastewater from the reactor vent gas scrubber in the production of ethylene dibromide via bromination of ethene.	(T)
K118	Spent adsorbent solids from purification of ethylene dibromide in the production of ethylene dibromide via bromination of ethene.	(T)
K136	Still bottoms from the purification of ethylene dibromide in the production of ethylene dibromide via bromination of ethene.	(T)
Inorganic chemicals:		
K071	Brine purification muds from the mercury cell process in chlorine production, where separately prepurified brine is not used.	(T)
K073	Chlorinated hydrocarbon waste from the purification step of the diaphragm cell process using graphite anodes in chlorine production.	(T)
K106	Wastewater treatment sludge from the mercury cell process in chlorine production	(T)
Pesticides:		
K031	By-product salts generated in the production of MSMA and cacodylic acid	(T)
K032	Wastewater treatment sludge from the production of chlordane	(T)
K033	Wastewater and scrub water from the chlorination of cyclopentadiene in the production of chlordane.	(T)
K034	Filter solids from the filtration of hexachlorocyclopentadiene in the production of chlordane.	(T)
K097	Vacuum stripper discharge from the chlordane chlorinator in the production of chlordane.	(T)
K035	Wastewater treatment sludges generated in the production of creosote	(T)
K036	Still bottoms from toluene reclamation distillation in the production of disulfoton	(T)
K037	Wastewater treatment sludges from the production of disulfoton	(T)
K038	Wastewater from the washing and stripping of phorate production	(T)
K039	Filter cake from the filtration of diethylphosphorodithioic acid in the production of phorate.	(T)

Environmental Protection Agency   § 261.33

Industry and EPA hazardous waste No.	Hazardous waste	Hazard code
K040	Wastewater treatment sludge from the production of phorate	(T)
K041	Wastewater treatment sludge from the production of toxaphene	(T)
K098	Untreated process wastewater from the production of toxaphene	(T)
K042	Heavy ends or distillation residues from the distillation of tetrachlorobenzene in the production of 2,4,5-T.	(T)
K043	2,6-Dichlorophenol waste from the production of 2,4-D	(T)
K099	Untreated wastewater from the production of 2,4-D	(T)
Explosives:		
K044	Wastewater treatment sludges from the manufacturing and processing of explosives ..	(R)
K045	Spent carbon from the treatment of wastewater containing explosives.........................	(R)
K046	Wastewater treatment sludges from the manufacturing, formulation and loading of lead-based initiating compounds.	(T)
K047	Pink/red water from TNT operations	(R)
Petroleum refining:		
K048	Dissolved air flotation (DAF) float from the petroleum refining industry.......................	(T)
K049	Slop oil emulsion solids from the petroleum refining industry......................................	(T)
K050	Heat exchanger bundle cleaning sludge from the petroleum refining industry................	(T)
K051	API separator sludge from the petroleum refining industry	(T)
K052	Tank bottoms (leaded) from the petroleum refining industry	(T)
Iron and steel:		
K061	Emission control dust/sludge from the primary production of steel in electric furnaces.	(T)
K062	Spent pickle liquor generated by steel finishing operations of plants that produce iron or steel.	(C,T)
Secondary lead:		
K069	Emission control dust/sludge from secondary lead smelting..	(T)
K100	Waste leaching solution from acid leaching of emission control dust/sludge from secondary lead smelting.	(T)
Veterinary pharmaceuticals:		
K084	Wastewater treatment sludges generated during the production of veterinary pharmaceuticals from arsenic or organo-arsenic compounds.	(T)
K101	Distillation tar residues from the distillation of aniline-based compounds in the production of veterinary pharmaceuticals from arsenic or organo-arsenic compounds.	(T)
K102	Residue from the use of activated carbon for decolorization in the production of veterinary pharmaceuticals from arsenic or organo-arsenic compounds.	(T)
Ink formulation: K086	Solvent washes and sludges, caustic washes and sludges, or water washes and sludges from cleaning tubs and equipment used in the formulation of ink from pigments, driers, soaps, and stabilizers containing chromium and lead.	(T)
Coking:		
K060	Ammonia still lime sludge from coking operations..	(T)
K087	Decanter tank tar sludge from coking operations...	(T)

§§261.33 Discarded commercial chemical products, off-specification species, container residues, and spill residues thereof.

The following materials or items are hazardous wastes if and when they are discarded or intended to be discarded, when they are mixed with waste oil or used oil or other material and applied to the land for dust suppression or road treatment, or when, in lieu of their original intended use, they are produced for use as (or as a component of) a fuel, distributed for use as a fuel, or burned as a fuel.

(a) Any commercial chemical product, or manufacturing chemical intermediate having the generic name listed in paragraph (e) or (f) of this section.

(b) Any off-specification commercial chemical product or manufacturing chemical intermediate which, if it met specifications, would have the generic name listed in paragraph (e) or (f) of this section.

(c) Any container or inner liner removed from a container that has been used to hold any commercial chemical product or manufacturing chemical intermediate having the generic names listed in paragraph (e) of this section, or any container or inner liner removed from a container that has been used to hold any off-specification chemical produce and manufacturing chemical intermediate which, if it met specifications would have the generic name listed in paragraph (e) of this section, unless the container is empty as defined in § 261.7(b)(3) of this chapter.

[*Comment*: Unless the residue is being beneficially used or reused, or legitimately recycled or reclaimed; or being accumulated, stored, transported or treated prior to such use, re-use recycling or reclamation, EPA considers the residue to be intended for discard, and thus a hazardous waste. An example of a legitimate re-use of the residue would be where the residue remains in the container and the container is used to hold the same commercial chemical product or manufacturing chemical product or manufacturing chemical product or manufacturing chemical intermediate if previously held. An example of the discard of the residue would be where the drum is sent to a drum reconditioner who reconditions the drum but discards the residue.]

(d) Any residue or contaminated soil, water or other debris resulting from the cleanup of a spill into or on any land or water of any commerical chemical product or manufacturing chemical intermediate having the generic name listed in paragraph (e) or (f) of this section, or any residue or contaminated soil, water or other debris resulting from the cleanup of a spill, into or on any land or water, of any off-specification chemical intermediate which, if it met specifications, would have the generic name listed in paragraph (e) or (f) of this section.

[*Comment*: The phrase "commercial chemical product or manufacturing chemical intermediate having the generic name listed in . . ." refers to a chemical substance which is manufactured or formulated for commercial or manufacturing use which consists of the commercially pure grade of the chemical any technical grades of the chemical that are produced or marketed, and all formulations in which the chemical is the sole active ingredient. It does not refer to a material, such as a manufacturing process waste, that contains any of the substances listed in paragraph (e) or (f). Where a manufacturing process waste is deemed to be a hazardous waste because it contains a substance listed in paragraph (e) or (f), such waste will be listed in either § 261.31 or § 261.32 or will be identified as a hazardous waste by the characteristics set forth in Subpart C of this part.]

(e) The commercial chemical products, manufacturing chemical intermediates or off-specification commercial chemical products or manufacturing chemical intermediates referred to in paragraphs (a) through (d) of this section, are identified as acute hazardous wastes (H) and are subject to be the small quantity exclusion defined in § 261.5(e).

[*Comment*: For the convenience of the regulated community the primary hazardous properties of these materials have been indicated by the letters T (Toxicity), and R (Reactivity). Absence of a letter indicates that the compound only is listed for acute toxicity.]

These wastes and their corresponding EPA Hazardous Waste Numbers are:

Hazardous waste No.	Substance	Hazardous waste No.	Substance
P023	Acetaldehyde, chloro-	P037	Dieldrin
P002	Acetamide, N-(aminothioxomethyl)-	P038	Diethylarsine
P057	Acetamide, 2-fluoro-	P039	O,O-Diethyl S-[2-(ethylthio)ethyl] phosphorodithioate
P058	Acetic acid, fluoro-, sodium salt		
P066	Acetimidic acid, N-[(methylcarbamoyl)oxy]thio-, methyl ester	P041	Diethyl-p-nitrophenyl phosphate
P001	3-(alpha-Acetonylbenzyl)-4-hydroxycoumarin and salts, when present at concentrations greater than 0.3%	P040	O,O-Diethyl O-pyrazinyl phosphorothioate
		P043	Diisopropyl fluorophosphate
		P044	Dimethoate
P002	1-Acetyl-2-thiourea	P045	3,3-Dimethyl-1-(methylthio)-2-butanone, O-[(methylamino)carbonyl] oxime
P003	Acrolein		
P070	Aldicarb	P071	O,O-Dimethyl O-p-nitrophenyl phosphorothioate
P004	Aldrin		
P005	Allyl alcohol	P082	Dimethylnitrosamine
P006	Aluminum phosphide	P046	alpha, alpha-Dimethylphenethylamine
P007	5-(Aminomethyl)-3-isoxazolol	P047	4,6-Dinitro-o-cresol and salts
P008	4-aAminopyridine	P034	4,6-Dinitro-o-cyclohexylphenol
P009	Ammonium picrate (R)	P048	2,4-Dinitrophenol
P119	Ammonium vanadate	P020	Dinoseb
P010	Arsenic acid	P085	Diphosphoramide, octamethyl-
P012	Arsenic (III) oxide	P039	Disulfoton
P011	Arsenic (V) oxide	P049	2,4-Dithiobiuret
P011	Arsenic pentoxide	P109	Dithiopyrophosphoric acid, tetraethyl ester
P012	Arsenic trioxide	P050	Endosulfan
P038	Arsine, diethyl-	P088	Endothall
P054	Aziridine	P051	Endrin
P013	Barium cyanide	P042	Epinephrine
P024	Benzenamine, 4-chloro-	P046	Ethanamine, 1,1-dimethyl-2-phenyl-
P077	Benzenamine, 4-nitro-	P084	Ethenamine, N-methyl-N-nitroso-
P028	Benzene, (chloromethyl)-	P101	Ethyl cyanide
P042	1,2-Benzenediol, 4-[1-hydroxy-2-(methylamino)ethyl]-	P054	Ethylenimine
		P097	Famphur
P014	Benzenethiol	P056	Fluorine
P028	Benzyl chloride	P057	Fluoroacetamide
P015	Beryllium dust	P058	Fluoroacetic acid, sodium salt
P016	Bis(chloromethyl) ether	P065	Fulminic acid, mercury(II) salt (R,T)
P017	Bromoacetone	P059	Heptachlor
P018	Brucine	P051	1,2,3,4,10,10-Hexachloro-6,7-epoxy-1,4,4a,5,6,7,8,8a-octahydro-endo,endo-1,4:5,8-dimethanonaphthalene
P021	Calcium cyanide		
P123	Camphene, octachloro-		
P103	Carbamimidoselenoic acid	P037	1,2,3,4,10,10-Hexachloro-6,7-epoxy-1,4,4a,5,6,7,8,8a-octahydro-endo,exo-1,4:5,8-demethanonaphthalene
P022	Carbon bisulfide		
P022	Carbon disulfide		
P095	Carbonyl chloride	P060	1,2,3,4,10,10-Hexachloro-1,4,4a,5,8,8a-hexahydro-1,4:5,8-endo, endo-dimeth- an-onaphthalene
P033	Chlorine cyanide		
P023	Chloroacetaldehyde		
P024	p-Chloroaniline	P004	1,2,3,4,10,10-Hexachloro-1,4,4a,5,8,8a-hexahydro-1,4:5,8-endo,exo-dimethanonaphthalene
P026	1-(o-Chlorophenyl)thiourea		
P027	3-Chloropropionitrile		
P029	Copper cyanides	P060	Hexachlorohexahydro-exo,exo-dimethanonaphthalene
P030	Cyanides (soluble cyanide salts), not elsewhere specified		
		P062	Hexaethyl tetraphosphate
P031	Cyanogen	P116	Hydrazinecarbothioamide
P033	Cyanogen chloride	P068	Hydrazine, methyl-
P036	Dichlorophenylarsine	P063	Hydrocyanic acid
		P063	Hydrogen cyanide

Hazardous waste No.	Substance	Hazardous waste No.	Substance
P096	Hydrogen phosphide	P110	Plumbane, tetraethyl-
P064	Isocyanic acid, methyl ester	P098	Potassium cyanide
P007	3(2H)-Isoxazolone, 5-(aminomethyl)-	P099	Potassium silver cyanide
P092	Mercury, (acetato-O)phenyl-	P070	Propanal, 2-methyl-2-(methylthio)-, O-[(methylamino)carbonyl]oxime
P065	Mercury fulminate (R,T)		
P016	Methane, oxybis(chloro-	P101	Propanenitrile
P112	Methane, tetranitro- (R)	P027	Propanenitrile, 3-chloro-
P118	Methanethiol, trichloro-	P069	Propanenitrile, 2-hydroxy-2-methyl-
P059	4,7-Methano-1H-indene, 1,4,5,6,7,8,8-heptachloro-3a,4,7,7a-tetrahydro-	P081	1,2,3-Propanetriol, trinitrate- (R)
		P017	2-Propanone, 1-bromo-
P066	Methomyl	P102	Propargyl alcohol
P067	2-Methylaziridine	P003	2-Propenal
P068	Methyl hydrazine	P005	2-Propen-1-ol
P064	Methyl isocyanate	P067	1,2-Propylenimine
P069	2-Methyllactonitrile	P102	2-Propyn-1-ol
P071	Methyl parathion	P008	4-Pyridinamine
P072	alpha-Naphthylthiourea	P075	Pyridine, (S)-3-(1-methyl-2-pyrrolidinyl)-, and salts
P073	Nickel carbonyl		
P074	Nickel cyanide	P111	Pyrophosphoric acid, tetraethyl ester
P074	Nickel(II) cyanide	P103	Selenourea
P073	Nickel tetracarbonyl	P104	Silver cyanide
P075	Nicotine and salts	P105	Sodium azide
P076	Nitric oxide	P106	Sodium cyanide
P077	p-Nitroaniline	P107	Strontium sulfide
P078	Nitrogen dioxide	P108	Strychnidin-10-one, and salts
P076	Nitrogen(II) oxide	P018	Strychnidin-10-one, 2,3-dimethoxy-
P078	Nitrogen(IV) oxide	P108	Strychnine and salts
P081	Nitroglycerine (R)	P115	Sulfuric acid, thallium(I) salt
P082	N-Nitrosodimethylamine	P109	Tetraethyldithiopyrophosphate
P084	N-Nitrosomethylvinylamine	P110	Tetraethyl lead
P050	5-Norbornene-2,3-dimethanol, 1,4,5,6,7,7-hexachloro, cyclic sulfite	P111	Tetraethylpyrophosphate
		P112	Tetranitromethane (R)
P085	Octamethylpyrophosphoramide	P062	Tetraphosphoric acid, hexaethyl ester
P087	Osmium oxide	P113	Thallic oxide
P087	Osmium tetroxide	P113	Thallium(III) oxide
P088	7-Oxabicyclo[2.2.1]heptane-2,3-dicarboxylic acid	P114	Thallium(I) selenite
		P115	Thallium(I) sulfate
P089	Parathion	P045	Thiofanox
P034	Phenol, 2-cyclohexyl-4,6-dinitro-	P049	Thioimidodicarbonic diamide
P048	Phenol, 2,4-dinitro-	P014	Thiophenol
P047	Phenol, 2,4-dinitro-6-methyl-	P116	Thiosemicarbazide
P020	Phenol, 2,4-dinitro-6-(1-methylpropyl)-	P026	Thiourea, (2-chlorophenyl)-
P009	Phenol, 2,4,6-trinitro-, ammonium salt (R)	P072	Thiourea, 1-naphthalenyl-
P036	Phenyl dichloroarsine	P093	Thiourea, phenyl-
P092	Phenylmercuric acetate	P123	Toxaphene
P093	N-Phenylthiourea	P118	Trichloromethanethiol
P094	Phorate	P119	Vanadic acid, ammonium salt
P095	Phosgene	P120	Vanadium pentoxide
P096	Phosphine	P120	Vanadium(V) oxide
P041	Phosphoric acid, diethyl p-nitrophenyl ester	P001	Warfarin, when present at concentrations greater than 0.3%
P044	Phosphorodithioic acid, O,O-dimethyl S-[2-(methylamino)-2-oxoethyl]ester		
		P121	Zinc cyanide
P043	Phosphorofluoric acid, bis(1-methylethyl)-ester	P122	Zinc phosphide (R,T)
		P122	Zinc phosphide, when present at concentrations greater than 10%
P094	Phosphorothioic acid, O,O-diethyl S-(ethylthio)methyl ester		
P089	Phosphorothioic acid, O,O-diethyl O-(p-nitrophenyl) ester		
P040	Phosphorothioic acid, O,O-diethyl O- pyrazinyl ester		
P097	Phosphorothioic acid, O,O-dimethyl O-[p-((dimethylamino)-sulfonyl)phenyl]ester		

(f) The commercial chemical products, manufacturing chemical intermediates, or off-specification commercial chemical products referred to in paragraphs (a) through (d) of this section, are identified as toxic wastes (T), unless otherwise designated and are subject to the small quantity generator exclusion defined in § 261.5 (a) and (g).

[*Comment*: For the convenience of the regulated community, the primary hazardous properties of these materials have been indicated by the letters T (Toxicity), R (Reactivity), I (Ignitability) and C (Corrosivity). Absence of a letter indicates that the compound is only listed for toxicity.]

These wastes and their corresponding EPA Hazardous Waste Numbers are:

Hazardous Waste No.	Substance	Hazardous Waste No.	Substance
U001	Acetaldehyde (I)	U069	1,2-Benzenedicarboxylic acid, dibutyl ester
U034	Acetaldehyde, trichloro-	U088	1,2-Benzenedicarboxylic acid, diethyl ester
U187	Acetamide, N-(4-ethoxyphenyl)-	U102	1,2-Benzenedicarboxylic acid, dimethyl ester
U005	Acetamide, N-9H-fluoren-2-yl-	U107	1,2-Benzenedicarboxylic acid, di-n-octyl ester
U112	Acetic acid, ethyl ester (I)	U070	Benzene, 1,2-dichloro-
U144	Acetic acid, lead salt	U071	Benzene, 1,3-dichloro-
U214	Acetic acid, thallium(I) salt	U072	Benzene, 1,4-dichloro-
U002	Acetone (I)	U017	Benzene, (dichloromethyl)-
U003	Acetonitrile (I,T)	U223	Benzene, 1,3-diisocyanatomethyl- (R,T)
U248	3-(alpha-Acetonylbenzyl)-4-hydroxycoumarin and salts, when present at concentrations of 0.3% or less	U239	Benzene, dimethyl-(I,T)
		U201	1,3-Benzenediol
		U127	Benzene, hexachloro-
U004	Acetophenone	U056	Benzene, hexahydro- (I)
U005	2-Acetylaminofluorene	U188	Benzene, hydroxy-
U006	Acetyl chloride (C,R,T)	U220	Benzene, methyl-
U007	Acrylamide	U105	Benzene, 1-methyl-1-2,4-dinitro-
U008	Acrylic acid (I)	U106	Benzene, 1-methyl-2,6-dinitro-
U009	Acrylonitrile	U203	Benzene, 1,2-methylenedioxy-4-allyl-
U150	Alanine, 3-[p-bis(2-chloroethyl)amino] phenyl-, L-	U141	Benzene, 1,2-methylenedioxy-4-propenyl-
		U090	Benzene, 1,2-methylenedioxy-4-propyl-
U328	2-Amino-I-methylbenzene	U055	Benzene, (1-methylethyl)- (I)
U353	4-Amino-I-methylbenzene	U169	Benzene, nitro- (I,T)
U011	Amitrole	U183	Benzene, pentachloro-
U012	Aniline (I,T)	U185	Benzene, pentachloro-nitro-
U014	Auramine	U020	Benzenesulfonic acid chloride (C,R)
U015	Azaserine	U020	Benzenesulfonyl chloride (C,R)
U010	Azirino(2′,3′:3,4)pyrrolo(1,2-a)indole-4,7-dione, 6-amino-8-[((aminocarbonyl) oxy)methyl]-1,1a,2,8,8a,8b-hexahydro-8a-methoxy-5-methyl-,	U207	Benzene, 1,2,4,5-tetrachloro-
		U023	Benzene, (trichloromethyl)-(C,R,T)
		0234	Benzene, 1,3,5-trinitro- (R,T)
		U021	Benzidine
U157	Benz[j]aceanthrylene, 1,2-dihydro-3-methyl-	U202	1,2-Benzisothiazolin-3-one, 1,1-dioxide
U016	Benz[c]acridine	U120	Benzo[j,k]fluorene
U016	3,4-Benzacridine	U022	Benzo[a]pyrene
U017	Benzal chloride	U022	3,4-Benzopyrene
U018	Benz[a]anthracene	U197	p-Benzoquinone
U018	1,2-Benzanthracene	U023	Benzotrichloride (C,R,T)
U094	1,2-Benzanthracene, 7,12-dimethyl-	U050	1,2-Benzphenanthrene
U012	Benzenamine (I,T)	U085	2,2′-Bioxirane (I,T)
U014	Benzenamine, 4,4′-carbonimidoylbis(N,N-di-methyl-	U021	(1,1′-Biphenyl)-4,4′-diamine
		U073	(1,1′-Biphenyl)-4,4′-diamine, 3,3′-dichloro-
U049	Benzenamine, 4-chloro-2-methyl-	U091	(1,1′-Biphenyl)-4,4′-diamine, 3,3′-dimethoxy-
U093	Benzenamine, N,N′-dimethyl-4-phenylazo-	U095	(1,1′-Biphenyl)-4,4′-diamine, 3,3′-dimethyl-
U158	Benzenamine, 4,4′-methylenebis(2-chloro-	U024	Bis(2-chloroethoxy) methane
U222	Benzenamine, 2-methyl-, hydrochloride	U027	Bis(2-chloroisopropyl) ether
U181	Benzenamine, 2-methyl-5-nitro-	U244	Bis(dimethylthiocarbamoyl) disulfide
U019	Benzene (I,T)	U028	Bis(2-ethylhexyl) phthalate
U038	Benzeneacetic acid, 4-chloro-alpha-(4-chloro-phenyl)-alpha-hydroxy, ethyl ester	U246	Bromine cyanide
		U225	Bromoform
U030	Benzene, 1-bromo-4-phenoxy-	U030	4-Bromophenyl phenyl ether
U037	Benzene, chloro-	U128	1,3-Butadiene, 1,1,2,3,4,4-hexachloro-
U190	1,2-Benzenedicarboxylic acid anhydride	U172	1-Butanamine, N-butyl-N-nitroso-
U028	1,2-Benzenedicarboxylic acid, [bis(2-ethyl-hexyl)] ester	U035	Butanoic acid, 4-[Bis(2-chloroethyl)amino] benzene-
		U031	1-Butanol (I)

Hazardous Waste No.	Substance
U159	2-Butanone (I,T)
U160	2-Butanone peroxide (R,T)
U053	2-Butenal
U074	2-Butene, 1,4-dichloro- (I,T)
U031	n-Butyl alchohol (I)
U136	Cacodylic acid
U032	Calcium chromate
U238	Carbamic acid, ethyl ester
U178	Carbamic acid, methylnitroso-, ethyl ester
U176	Carbamide, N-ethyl-N-nitroso-
U177	Carbamide, N-methyl-N-nitroso-
U219	Carbamide, thio-
U097	Carbamoyl chloride, dimethyl-
U215	Carbonic acid, dithallium(I) salt
U156	Carbonochloridic acid, methyl ester (I,T)
U033	Carbon oxyfluoride (R,T)
U211	Carbon tetrachloride
U033	Carbonyl fluoride (R,T)
U034	Chloral
U035	Chlorambucil
U036	Chlordane, technical
U026	Chlornaphazine
U037	Chlorobenzene
U039	4-Chloro-m-cresol
U041	1-Chloro-2,3-epoxypropane
U042	2-Chloroethyl vinyl ether
U044	Chloroform
U046	Chloromethyl methyl ether
U047	beta-Chloronaphthalene
U048	o-Chlorophenol
U049	4-Chloro-o-toluidine, hydrochloride
U032	Chromic acid, calcium salt
U050	Chrysene
U051	Creosote
U052	Cresols
U052	Cresylic acid
U053	Crotonaldehyde
U055	Cumene (I)
U246	Cyanogen bromide
U197	1,4-Cyclohexadienedione
U056	Cyclohexane (I)
U057	Cyclohexanone (I)
U130	1,3-Cyclopentadiene, 1,2,3,4,5,5-hexa- chloro-
U058	Cyclophosphamide
U240	2,44-D, salts and esters
U059	Daunomycin
U060	DDD
U061	DDT
U142	Decachlorooctahydro-1,3,4-metheno-2H-cyclobuta[c,d]-pentalen-2-one
U062	Diallate
U133	Diamine (R,T)
U221	Diaminotoluene
U063	Dibenz[a,h]anthracene
U063	1,2:5,6-Dibenzanthracene
U064	1,2:7,8-Dibenzopyrene
U064	Dibenz[a,i]pyrene
U066	1,2-Dibromo-3-chloropropane
U069	Dibutyl phthalate
U062	S-(2,3-Dichloroallyl) diisopropylthiocarbamate
U070	o-Dichlorobenzene
U071	m-Dichlorobenzene
U072	p-Dichlorobenzene
U073	3,3'-Dichlorobenzidine
U074	1,4-Dichloro-2-butene (I,T)
U075	Dichlorodifluoromethane
U192	3,5-Dichloro-N-(1,1-dimethyl-2-propynyl) benzamide
U060	Dichloro diphenyl dichloroethane
U061	Dichloro diphenyl trichloroethane
U078	1,1-Dichloroethylene
U079	1,2-Dichloroethylene
U025	Dichloroethyl ether
U081	2,4-Dichlorophenol
U082	2,6-Dichlorophenol
U240	2,4-Dichlorophenoxyacetic acid, salts and esters
U083	1,2-Dichloropropane
U084	1,3-Dichloropropene
U085	1,2:3,4-Diepoxybutane (I,T)
U108	1,4-Diethylene dioxide

Hazardous Waste No.	Substance
U086	N,N-Diethylhydrazine
U087	O,O-Diethyl-S-methyl-dithiophosphate
U088	Diethyl phthalate
U089	Diethylstilbestrol
U148	1,2-Dihydro-3,6-pyradizinedione
U090	Dihydrosafrole
U091	3,3'-Dimethoxybenzidine
U092	Dimethylamine (I)
U093	Dimethylaminoazobenzene
U094	7,12-Dimethylbenz[a]anthracene
U095	3,3'-Dimethylbenzidine
U096	alpha,alpha-Dimethylbenzylhydroperoxide (R)
U097	Dimethylcarbamoyl chloride
U098	1,1-Dimethylhydrazine
U099	1,2-Dimethylhydrazine
U101	2,4-Dimethylphenol
U102	Dimethyl phthalate
U103	Dimethyl sulfate
U105	2,4-Dinitrotoluene
U106	2,6-Dinitrotoluene
U107	Di-n-octyl phthalate
U108	1,4-Dioxane
U109	1,2- Diphenylhydrazine
U110	Dipropylamine (I)
U111	Di-N-propylnitrosamine
U001	Ethanal (I)
U174	Ethanamine, N-ethyl-N-nitroso-
U067	Ethane, 1,2-dibromo-
U076	Ethane, 1,1-dichloro-
U077	Ethane, 1,2-dichloro-
U114	1,2-Ethanediylbiscarbamodithioic acid
U131	Ethane, 1,1,1,2,2,2-hexachloro-
U024	Ethane, 1,1'-[methylenebis(oxy)]bis[2-chloro-
U003	Ethanenitrile (I, T)
U117	Ethane,1,1'-oxybis- (I)
U025	Ethane, 1,1'-oxybis[2-chloro-
U184	Ethane, pentachloro-
U208	Ethane, 1,1,1,2-tetrachloro-
U209	Ethane, 1,1,2,2-tetrachloro-
U218	Ethanethioamide
U247	Ethane, 1,1,1,-trichloro-2,2-bis(p-methoxyphenyl).
U227	Ethane, 1,1,2-trichloro-
U043	Ethene, chloro-
U042	Ethene, 2-chloroethoxy-
U078	Ethene, 1,1-dichloro-
U079	Ethene, trans-1,2-dichloro-
U210	Ethene, 1,1,2,2-tetrachloro-
U173	Ethanol, 2,2'-(nitrosoimino)bis-
U004	Ethanone, 1-phenyl-
U006	Ethanoyl chloride (C,R,T)
U359	2-Ethoxyethanol
U112	Ethyl acetate (I)
U113	Ethyl acrylate (I)
U238	Ethyl carbamate (urethan)
U038	Ethyl 4,4'-dichlorobenzilate
U359	Ethylene glycol monoethyl ether.
U114	Ethylenebis(dithiocarbamic acid)
U067	Ethylene dibromide
U077	Ethylene dichloride
U115	Ethlene oxide (I,T)
U116	Ethylene thiourea
U117	Ethyl ether (I)
U076	Ethylidene dichloride
U118	Ethylmethacrylate
U119	Ethyl methanesulfonate
U139	Ferric dextran
U120	Fluoranthene
U122	Formaldehyde
U123	Formic acid (C,T)
U124	Furan (I)
U125	2-Furancarboxaldehyde (I)
U147	2,5-Furandione
U213	Furan, tetrahydro- (I)
U125	Furfural (I)
U124	Furfuran (I)
U206	D-Glucopyranose, 2-deoxy-2(3-methyl-3-nitro-soureido)-
U126	Glycidylaldehyde
U163	Guanidine, N-nitroso-N-methyl-N'-nitro-

Hazardous Waste No.	Substance
U127	Hexachlorobenzene
U128	Hexachlorobutadiene
U129	Hexachlorocyclohexane (gamma isomer)
U130	Hexachlorocyclopentadiene
U131	Hexachloroethane
U132	Hexachlorophene
U243	Hexachloropropene
U133	Hydrazine (R,T)
U086	Hydrazine, 1,2-diethyl-
U098	Hydrazine, 1,1-dimethyl-
U099	Hydrazine, 1,2-dimethyl-
U109	Hydrazine, 1,2-diphenyl-
U134	Hydrofluoric acid (C,T)
U134	Hydrogen fluoride (C,T)
U135	Hydrogen sulfide
U096	Hydroperoxide, 1-methyl-1-phenylethyl- (R)
U136	Hydroxydimethylarsine oxide
U116	2-Imidazolidinethione
U137	Indeno[1,2,3-cd]pyrene
U139	Iron dextran
U140	Isobutyl alcohol (I,T)
U141	Isosafrole
U142	Kepone
U143	Lasiocarpine
U144	Lead acetate
U145	Lead phosphate
U146	Lead subacetate
U129	Lindane
U147	Maleic anhydride
U148	Maleic hydrazide
U149	Malononitrile
U150	Melphalan
U151	Mercury
U152	Methacrylonitrile (I,T)
U092	Methanamine, N-methyl- (I)
U029	Methane, bromo-
U045	Methane, chloro- (I,T)
U046	Methane, chloromethoxy-
U068	Methane, dibromo-
U080	Methane, dichloro-
U075	Methane, dichlorodifluoro-
U138	Methane, iodo-
U119	Methanesulfonic acid, ethyl ester
U211	Methane, tetrachloro-
U121	Methane, trichlorofluoro-
U153	Methanethiol (I,T)
U225	Methane, tribromo-
U044	Methane, trichloro-
U121	Methane, trichlorofluoro-
U123	Methanoic acid (C,T)
U036	4,7-Methanoindan, 1,2,4,5,6,7,8,8-octachloro-3a,4,7,7a-tetrahydro-
U154	Methanol (I)
U155	Methapyrilene
U247	Methoxychlor.
U154	Methyl alcohol (I)
U029	Methyl bromide
U186	1-Methylbutadiene (I)
U045	Methyl chloride (I,T)
U156	Methyl chlorocarbonate (I,T)
U226	Methylchloroform
U157	3-Methylcholanthrene
U158	4,4'-Methylenebis(2-chloroaniline)
U132	2,2'-Methylenebis(3,4,6-trichlorophenol)
U068	Methylene bromide
U080	Methylene chloride
U122	Methylene oxide
U159	Methyl ethyl ketone (I,T)
U160	Methyl ethyl ketone peroxide (R,T)
U138	Methyl iodide
U161	Methyl isobutyl ketone (I)
U162	Methyl methacrylate (I,T)
U163	N-Methyl-N'-nitro-N-nitrosoguanidine
U161	4-Methyl-2-pentanone (I)
U164	Methylthiouracil
U010	Mitomycin C
U059	5,12-Naphthacenedione, (8S-cis)-8-acetyl-10-[(3-amino-2,3,6-trideoxy-alpha-L-lyxo-hexopyranosyl)oxy]-7,8,9,10-tetrahydro-6,8,11-trihydroxy-1-methoxy-
U165	Naphthalene
U047	Naphthalene, 2-chloro-
U166	1,4-Naphthalenedione

Hazardous Waste No.	Substance
U236	2,7-Naphthalenedisulfonic acid, 3,3'-[(3,3'-dimethyl-(1,1'-biphenyl)-4,4'diyl)]-bis(azo)bis(5-amino-4-hydroxy)-,tetrasodium salt
U166	1,4,Naphthaquinone
U167	1-Naphthylamine
U168	2-Naphthylamine
U167	alpha-Naphthylamine
U168	beta-Naphthylamine
U026	2-Naphthylamine, N,N'-bis(2-chloromethyl)-
U169	Nitrobenzene (I,T)
U170	p-Nitrophenol
U171	2-Nitropropane (I,T)
U172	N-Nitrosodi-n-butylamine
U173	N-Nitrosodiethanolamine
U174	N-Nitrosodiethylamine
U111	N-Nitroso-N-propylamine
U176	N-Nitroso-N-ethylurea
U177	N-Nitroso-N-methylurea
U178	N-Nitroso-N-methylurethane
U179	N-Nitrosopiperidine
U180	N-Nitrosopyrrolidine
U181	5-Nitro-o-toluidine
U193	1,2-Oxathiolane, 2,2-dioxide
U058	2H-1,3,2-Oxazaphosphorine, 2-[bis(2-chloroethyl)amino]tetrahydro-, oxide 2-
U115	Oxirane (I,T)
U041	Oxirane, 2-(chloromethyl)-
U182	Paraldehyde
U183	Pentachlorobenzene
U184	Pentachloroethane
U185	Pentachloronitrobenzene
See F027	Pentachlorophenol
U186	1,3-Pentadiene (I)
U187	Phenacetin
U188	Phenol
U048	Phenol, 2-chloro-
U039	Phenol, 4-chloro-3-methyl-
U081	Phenol, 2,4-dichloro-
U082	Phenol, 2,6-dichloro-
U101	Phenol, 2,4-dimethyl-
U170	Phenol, 4-nitro-
See F027	Phenol, pentachloro-
Do	Phenol, 2,3,4,6-tetrachloro-
Do	Phenol, 2,4,5-trichloro-
Do	Phenol, 2,4,6-trichloro-
U137	1,10-(1,2-phenylene)pyrene
U145	Phosphoric acid, Lead salt
U087	Phosphorodithioic acid, O,O-diethyl-, S-methyl-ester
U189	Phosphorous sulfide (R)
U190	Phthalic anhydride
U191	2-Picoline
U192	Pronamide
U194	1-Propanamine (I,T)
U110	1-Propanamine, N-propyl- (I)
U066	Propane, 1,2-dibromo-3-chloro-
U149	Propanedinitrile
U171	Propane, 2-nitro- (I,T)
U027	Propane, 2,2'oxybis[2-chloro-
U193	1,3-Propane sultone
U235	1-Propanol, 2,3-dibromo-, phosphate (3:1)
U126	1-Propanol, 2,3-epoxy-
U140	1-Propanol, 2-methyl- (I,T)
U002	2-Propanone (I)
U007	2-Propanamide
U084	Propene, 1,3-dichloro-
U243	1-Propene, 1,1,2,3,3,3-hexachloro-
U009	2-Propenenitrile
U152	2-Propenenitrile, 2-methyl- (I,T)
U008	2-Propenoic acid (I)
U113	2-Propenoic acid, ethyl ester (I)
U118	2-Propenoic acid, 2-methyl-, ethyl ester
U162	2-Propenoic acid, 2-methyl-, methyl ester (I,T)
See F027	Propionic acid, 2-(2,4,5-trichlorophenoxy)-
U194	n-Propylamine (I,T)
U083	Propylene dichloride
U196	Pyridine
U155	Pyridine, 2-[(2-(dimethylamino)-2-thenylamino]-
U179	Pyridine, hexahydro-N-nitroso-

Hazardous Waste No.	Substance
U191	Pyridine, 2-methyl-
U164	4(1H)-Pyrimidinone, 2,3-dihydro-6-methyl-2-thioxo-
U180	Pyrrole, tetrahydro-N-nitroso-
U200	Reserpine
U201	Resorcinol
U202	Saccharin and salts
U203	Safrole
U204	Selenious acid
U204	Selenium dioxide
U205	Selenium disulfide (R,T)
U015	L-Serine, diazoacetate (ester)
See F027	Silvex
U089	4,4'-Stilbenediol, alpha,alpha'-diethyl-
U206	Streptozotocin
U135	Sulfur hydride
U103	Sulfuric acid, dimethyl ester
U189	Sulfur phosphide (R)
U205	Sulfur selenide (R,T)
See F027	2,4,5-T
U207	1,2,4,5-Tetrachlorobenzene
U208	1,1,1,2-Tetrachloroethane
U209	1,1,2,2-Tetrachloroethane
U210	Tetrachloroethylene
See F027	2,3,4,6-Tetrachlorophenol
U213	Tetrahydrofuran (I)
U214	Thallium(I) acetate
U215	Thallium(I) carbonate
U216	Thallium(I) chloride
U217	Thallium(I) nitrate
U218	Thioacetamide
U153	Thiomethanol (I,T)
U219	Thiourea
U244	Thiram
U220	Toluene
U221	Toluenediamine
U223	Toluene diisocyanate (R,T)
U328	o-Toluidine
U222	O-Toluidine hydrochloride
U353	p-Toluidine
U011	1H-1,2,4-Triazol-3-amine
U226	1,1,1-Trichloroethane
U227	1,1,2-Trichloroethane
U228	Trichloroethene
U228	Trichloroethylene
U121	Trichloromonofluoromethane
See F027	2,4,5-Trichlorophenol
Do	2,4,6-Trichlorophenol
Do	2,4,5-Trichlorophenoxyacetic acid
U234	sym-Trinitrobenzene (R,T)
U182	1,3,5-Trioxane, 2,4,5-trimethyl-
U235	Tris(2,3-dibromopropyl) phosphate
U236	Trypan blue
U237	Uracil, 5[bis(2-chloromethyl)amino]-
U237	Uracil mustard
U043	Vinyl chloride
U248	Warfarin, when present at concentrations of 0.3% or less
U239	Xylene (I)
U200	Yohimban-16-carboxylic acid, 11,17-dimethoxy-18-[(3,4,5-trimethoxy-benzoyl)oxy]-, methyl ester
U249	Zinc phosphide, when present at concentrations of 10% or less.

APPENDIX V

Table 1—Maximum Concentration of Contaminants for
Characteristic of EP Toxicity

EPA Hazardous Waste Number	Contaminant	Maximum Concentration (mg/L)
D004	Arsenic	5.0
D005	Barium	100.0
D006	Cadmium	1.0
D007	Chromium	5.0
D008	Lead	5.0
D009	Mercury	0.2
D010	Selenium	1.0
D011	Silver	5.0
D012	Endrin (1,2,3,4,10,10-hexachloro-1,7-epoxy-1,4,4a,5,6,7,8,8a-octahydro-1,4-endo, endo-5,8-dimethano-naphthalene)	0.02
D013	Lindane (1,2,3,4,5,6-hexachlorocyclohexane, gamma isomer)	0.4
D014	Methoxychlor (1,1,1-Trichloro-2,2-bis [p-methoxyphenyl]ethane)	10.0
D015	Toxaphene ($C_{10} H_{10} Cl_8$, Technical chlorinated camphene, 67–69 percent chlorine)	0.5
D016	2,4-D (2,4-Dichlorophenoxyacetic acid)	10.0
D017	2,4,5-TP Silvex (2,4,5-Trichlorophenoxypropionic acid)	1.0

Source: 40 CFR 261.24.

APPENDIX VI

Instructions for Completing the Uniform Hazardous Waste Manifest

Source: Reprinted from 40 CFR Part 262 Appendix.

U.S. EPA Form 8700-22

Read all instructions before completing this form.

This form has been designed for use on a 12-pitch (elite) typewriter; a firm point pen may also be used—press down hard.

Federal regulations require generators and transporters of hazardous waste and owners or operators of hazardous waste treatment, storage, and disposal facilities to use this form (8700-22) and, if necessary, the continuation sheet (Form 8700-22A) for both inter and intrastate transportation.

Federal regulations also require generators and transporters of hazardous waste and owners or operators of hazardous waste treatment, storage and disposal facilities to complete the following information:

GENERATORS

*Item 1. Generator's U.S. EPA ID Number—
Manifest Document Number*

Enter the generator's U.S. EPA twelve digit identification number and the unique five digit number assigned to this Manifest (e.g., 00001) by the generator.

Item 2. Page 1 of ——

Enter the total number of pages used to complete this Manifest, i.e., the first page (EPA Form 8700-22) plus the number of Continuation Sheets (EPA Form 8700-22A), if any.

*Item 3. Generator's Name and Mailing
Address*

Enter the name and mailing address of the generator. The address should be the location that will manage the returned Manifest forms.

Item 4. Generator's Phone Number

Enter a telephone number where an authorized agent' of the generator may be reached in the event of an emergency.

Item 5. Transporter 1 Company Name

Enter the company name of the first transporter who will transport the waste.

Item 6. U.S. EPA ID Number

Enter the U.S. EPA twelve digit identification number of the first transporter identified in item 5.

Item 7. Transporter 2 Company Name

If applicable, enter the company name of the second transporter who will transport the waste. If more than two transporters are used to transport the waste, use a Continuation Sheet(s) (EPA Form 8700-22A) and list the transporters in the order they will be transporting the waste.

Figure 1. EPA Form 8700–22, "Uniform Hazardous Waste Manifest."

Item 8. U.S. EPA ID Number

If applicable, enter the U.S. EPA twelve digit identification number of the second transporter identified in item 7.

NOTE: If more than two transporters are used, enter each additional transporter's company name and U.S. EPA twelve digit identification number in items 24-27 on the Continuation Sheet (EPA Form 8700-22A). Each Continuation Sheet has space to record two additional transporters. Every transporter used between the generator and the designated facility must be listed.

Item 9. Designated Facility Name and Site Address

Enter the company name and site address of the facility designated to receive the waste listed on this Manifest. The address must be the site address, which may differ from the company mailing address.

Item 10. U.S. EPA ID Number

Enter the U.S. EPA twelve digit identification number of the designated facility identified in item 9.

Item 11. U.S. DOT Description [Including Proper Shipping Name, Hazard Class, and ID Number (UN/NA)]

Enter the U.S. DOT Proper Shipping Name, Hazard Class, and ID Number (UN/NA) for each waste as identified in 49 CFR 171 through 177.

NOTE: If additional space is needed for waste descriptions, enter these additional descriptions in item 28 on the Continuation Sheet (EPA Form 8700-22A).

Item 12. Containers (No. and Type)

Enter the number of containers for each waste and the appropriate abbreviation from Table I (below) for the type of container.

Table I—Types of Containers

DM = Metal drums, barrels, kegs
DW = Wooden drums, barrels, kegs
DF = Fiberboard or plastic drums, barrels, kegs
TP = Tanks portable
TT = Cargo tanks (tank trucks)
TC = Tank cars
DT = Dump truck
CY = Cylinders
CM = Metal boxes, cartons, cases (including roll-offs)
CW = Wooden boxes, cartons, cases
CF = Fiber or plastic boxes, cartons, cases
BA = Burlap, cloth, paper or plastic bags

Item 13. Total Quantity

Enter the total quantity of waste described on each line.

Item 14. Unit (Wt./Vol.)

Enter the appropriate abbreviation from Table II (below) for the unit of measure.

Table II—Units of Measure

G = Gallons (liquids only)
P = Pounds
T = Tons (2000 lbs)
Y = Cubic yards
L = Liters (liquids only)
K = Kilograms
M = Metric tons (1000 kg)
N = Cubic meters

Item 15. Special Handling Instructions and Additional Information

Generators may use this space to indicate special transportation, treatment, storage, or disposal information or Bill of Lading information. States may not require additional, new, or different information in this space. For international shipments, generators must enter in this space the point of departure (City and State) for those shipments destined for treatment, storage, or disposal outside the jurisdiction of the United States.

Item 16. Generator's Certification

The generator must read, sign (by hand), and date the certification statement. If a mode *other than* highway is used, the word "highway" should be lined out and the appropriate mode (rail, water, or air) inserted in the space below. If another mode *in addition to* the highway mode is used, enter the appropriate additional mode (e.g., *and rail*) in the space below.

In signing the waste minimization certification statement, those generators who have not been exempted by statute or regulation from the duty to make a waste minimization certification under section 3002(b) of RCRA are also certifying that they have complied with the waste minimization requirements.

NOTE: All of the above information *except* the handwritten signature required in item 16 may be preprinted.

*   *   *   *   *

TRANSPORTERS

Item 17. Transporter 1 Acknowledgement of Receipt of Materials

Enter the name of the person accepting the waste on behalf of the first transporter. That person must acknowledge acceptance of the waste described on the Manifest by signing and entering the date of receipt.

Item 18. Transporter 2 Acknowledgement of Receipt of Materials

Enter, if applicable, the name of the person accepting the waste on behalf of the second transporter. That person must acknowledge acceptance of the waste described on the Manifest by signing and entering the date of receipt.

NOTE: International Shipments—Transporter Responsibilities.

Exports—Transporters must sign and enter the date the waste left the United States in Item 15 of Form 8700-22.

Imports—Shipments of hazardous waste regulated by RCRA and transported into the United States from another country must upon entry be accompanied by the U.S. EPA Uniform Hazardous Waste Manifest. Transporters who transport hazardous waste into the United States from another country are responsible for completing the Manifest (40 CFR 263.10(c)(1)).

Owners and Operators of Treatment, Storage, or Disposal Facilities

Item 19. Discrepancy Indication Space

The authorized representative of the designated (or alternate) facility's owner or operator must note in this space any significant discrepancy between the waste described on the Manifest and the waste actually received at the facility.

Owners and operators of facilities located in unauthorized States (i.e., the U.S. EPA administers the hazardous waste management program) who cannot resolve significant discrepancies within 15 days of receiving the waste must submit to their Regional Administrator (see list below) a letter with a copy of the Manifest at issue describing the discrepancy and attempts to reconcile it (40 CFR 264.72 and 265.72).

Owners and operators of facilities located in authorized States (i.e., those States that have received authorization from the U.S. EPA to administer the hazardous waste program) should contact their State agency for information on State Discrepancy Report requirements.

EPA Regional Administrators

Regional Administrator, U.S. EPA Region I, J.F. Kennedy Fed. Bldg., Boston, MA 02203

Regional Administrator, U.S. EPA Region II, 26 Federal Plaza, New York, NY 10278

Regional Administrator, U.S. EPA Region III, 6th and Walnut Sts., Philadelphia, PA 19106

Regional Administrator, U.S. EPA Region IV, 345 Courtland St., NE., Atlanta, GA 30365

Regional Administrator, U.S. EPA Region V, 230 S. Dearborn St., Chicago, IL 60604

Regional Administrator, U.S. EPA Region VI, 1201 Elm Street, Dallas, TX 75270

Regional Administrator, U.S. EPA Region VII, 324 East 11th Street, Kansas City, MO 64106

Regional Administrator, U.S. EPA Region VIII, 1860 Lincoln Street, Denver, CO 80295

Regional Administrator, U.S. EPA Region IX, 215 Freemont Street, San Francisco, CA 94105

Regional Administrator, U.S. EPA Region X, 1200 Sixth Avenue, Seattle, WA 98101

Item 20. Facility Owner or Operator: Certification of Receipt of Hazardous Materials Covered by This Manifest Except as Noted in Item 19

Print or type the name of the person accepting the waste on behalf of the owner or operator of the facility. That person must acknowledge acceptance of the waste described on the Manifest by signing and entering the date of receipt.

Items A-K are not required by Federal regulations for intra- or interstate transportation. However, States may require generators and owners or operators of treatment, storage, or disposal facilities to complete some or all of items A-K as part of State manifest reporting requirements. Generators and owners and operators of treatment, storage, or disposal facilities are advised to contact State officials for guidance on completing the shaded areas of the Manifest.

INSTRUCTIONS—CONTINUATION SHEET, U.S. EPA FORM 8700-22A

Read all instructions before completing this form.

This form has been designed for use on a 12-pitch (elite) typewriter; a firm point pen may also be used—press down hard.

This form must be used as a continuation sheet to U.S. EPA Form 8700-22 if:

• More than two transporters are to be used to transport the waste;
• More space is required for the U.S. DOT description and related information in Item 11 of U.S. EPA Form 8700-22.

Federal regulations require generators and transporters of hazardous waste and owners or operators of hazardous waste treatment, storage, or disposal facilities to use the uniform hazardous waste manifest (EPA Form 8700-22) and, if necessary, this continuation sheet (EPA Form 8700-22A) for both inter- and intrastate transportation.

GENERATORS

Item 21. Generator's U.S. EPA ID Number— Manifest Document Number

Enter the generator's U.S. EPA twelve digit identification number and the unique five digit number assigned to this Manifest (e.g., 00001) as it appears in item 1 on the first page of the Manifest.

Item 22. Page ——

Enter the page number of this Continuation Sheet.

Figure 2. EPA Form 8700–22A, "Uniform Hazardous Waste Manifest (Continuation Sheet)."

Item 23. Generator's Name

Enter the generator's name as it appears in item 3 on the first page of the Manifest.

Item 24. Transporter —— Company Name

If additional transporters are used to transport the waste described on this Manifest, enter the company name of each additional transporter in the order in which they will transport the waste. Enter after the word "Transporter" the order of the transporter. For example, Transporter *3* Company Name. Each Continuation Sheet will record the names of two additional transporters.

Item 25. U.S. EPA ID Number

Enter the U.S. EPA twelve digit identification number of the transporter described in item 24.

Item 26. Transporter —— Company Name

If additional transporters are used to transport the waste described on this Manifest, enter the company name of each additional transporter in the order in which they will transport the waste. Enter after the word "Transporter" the order of the transporter. For example, Transporter 4 Company Name. Each Continuation Sheet will record the names of two additional transporters.

Item 27. U.S. EPA ID Number

Enter the U.S. EPA twelve digit identification number of the transporter described in item 26.

Item 28. U.S. DOT Description Including Proper Shipping Name, Hazardous Class, and ID Number (UN/NA)

Refer to item 11.

Item 29. Containers (No. and Type)

Refer to item 12.

Item 30. Total Quantity

Refer to item 13.

Item 31. Unit (Wt./Vol.)

Refer to item 14.

Item 32. Special Handling Instructions

Generators may use this space to indicate special transportation, treatment, storage, or disposal information or Bill of Lading information. States are *not* authorized to require additional, new, or different information in this space.

●   ●   ●   ●   ●

TRANSPORTERS

Item 33. Transporter —— Acknowledgement of Receipt of Materials

Enter the same number of the Transporter as identified in item 24. Enter also the name of the person accepting the waste on behalf of the Transporter (Company Name) identified in item 24. That person must acknowledge acceptance of the waste described on the Manifest by signing and entering the date of receipt.

Item 34. Transporter —— Acknowledgement of Receipt of Materials

Enter the same number as identified in item 26. Enter also the name of the person accepting the waste on behalf of the Transporter (Company Name) identified in item 26. That person must acknowledge acceptance of the waste described on the Manifest by signing and entering the date of receipt.

●   ●   ●   ●   ●

OWNERS AND OPERATORS OF TREATMENT, STORAGE, OR DISPOSAL FACILITIES

Item 35. Discrepancy Indication Space

Refer to item 19.

Items L–R are not required by Federal regulations for intra- or interstate transportation. However, States may require generators and owners or operators of treatment, storage, or disposal facilities to complete some or all of items L–R as part of State manifest reporting requirements. Generators and owners and operators of treatment, storage, or disposal facilities are advised to contact State officials for guidance on completing the shaded areas of the manifest.

[49 FR 10501, Mar. 20, 1984, as amended at 50 FR 28745, July 15, 1985]

APPENDIX VII

U.S. DOT Hazardous Materials Definitions

Source: Courtesy of Information Services Division, Materials Transportation Bureau, United States Department of Transportation.

The following definitions have been abstracted from the Code of Federal Regulations, Title 49-Transportation, Parts 100-177. Refer to the referenced sections for complete details. NOTE: The plus (+) in Column (1), Sec. 172.101, Hazardous Materials Table) fixes the proper shipping name and hazard class for that entry without regard to whether the material meets the definition of that class. [Sec. 172.101(a)(1)]

HAZARDOUS MATERIAL – A substance or material which has been determined by the Secretary of Transportation to be capable of posing an unreasonable risk to health, safety and property when transported in commerce, and which has been so designated. (Sec. 171.8).

MULTIPLE HAZARDS – A material meeting the definition of more than one hazard class is classed according to the provisions set forth in Sec. 173.2(a) and (b).

HAZARD CLASS	UN No	DEFINITIONS
		An Explosive – Any chemical compound, mixture, or device, the primary or common purpose of which is to function by explosion, i.e., with substantially instantaneous release of gas and heat, unless such compound, mixture, or device is otherwise specifically classified in Parts 171-179. (Sec. 173.50)
CLASS A EXPLOSIVE	1	Detonating or otherwise of maximum hazard. The nine types of Class A explosives are defined in Sec. 173.53.
CLASS B EXPLOSIVE	1	In general, function by rapid combustion rather than detonation and include some explosive devices such as special fireworks, flash powders, etc. (Sec. 173.88)
CLASS C EXPLOSIVE	1	Certain types of manufactured articles containing Class A or Class B explosives, or both, as components but in restricted quantities, and certain types of fireworks. (Sec. 173.100)
BLASTING AGENT	1	A material designed for blasting which has been tested in accordance with Sec. 173.114a(b) and found to be so insensitive that there is very little probability of accidental initiation to explosion or of transition from deflagration to detonation. [Sec. 173.114a(b)]
		Compressed Gas – Any material or mixture having in the container a pressure EXCEEDING 40 psia at 70°F., or a pressure exceeding 104 psia at 130°F.; or any liquid flammable material having a vapor pressure exceeding 40 psia at 100°F. [Sec. 173.300(a)]
		Non-liquefied compressed gas is a gas, other than gas in solution, which under the charged pressure is entirely gaseous at a temperature of 70°F.
		Liquefied compressed gas is a gas which, under the charged pressure, is partially liquid at a temperature of 70°F.
		Compressed gas in solution is a nonliquefied compressed gas which is dissolved in a solvent.
FLAMMABLE GAS	2	Any compressed gas meeting the requirements for lower flammability limit, flammability limit range, flame projection, or flame propagation criteria as specified in Sec. 173.300(b).
NONFLAMMABLE GAS	2	Any compressed gas other than a flammable compressed gas.

HAZARD CLASS	UN No	DEFINITIONS
COMBUSTIBLE LIQUID	3	Any liquid having a flash point at or above 100°F. and below 200°F. as determined by tests listed in Sec. 173.115(d). Exceptions are found in Sec. 173.115(b).
FLAMMABLE LIQUID	3	Any liquid having a flash point below 100°F. as determined by tests listed in Sec. 173.115(d). For exceptions, see Sec. 173.115(a). Pyroforic Liquid – Any liquid that ignites spontaneously in dry or moist air at or below 130°F. [Sec. 173.115(c)]
FLAMMABLE SOLID	4	Any solid material, other than an explosive, which is liable to cause fires through friction, retained heat from manufacturing or processing, or which can be ignited readily and when ignited, burns so vigorously and persistently as to create a serious transportation hazard. Included in this class are spontaneously combustible and water-reactive materials. (Sec. 173.150) Spontaneously Combustible Material (Solid) – A solid substance (including sludges and pastes) which may undergo spontaneous heating or self-ignition under conditions normally incident to transportation or which may, upon contact with the atmosphere, undergo an increase in temperature and ignite. (Sec. 171.8) Water Reactive Material (Solid) – Any solid substance (including sludges and pastes) which, by interaction with water, is likely to become spontaneously flammable or to give off flammable or toxic gases in dangerous quantities. (Sec. 171.8)
ORGANIC PEROXIDE	5	An organic compound containing the bivalent –0-0 structure and which may be considered a derivative of hydrogen peroxide where one or more of the hydrogen atoms have been replaced by organic radicals must be classed as an organic peroxide unless...[See Sec. 173.151(a) for details].
OXIDIZER	5	A substance such as chlorate, permanganate, inorganic peroxide, or a nitrate, that yields oxygen readily to stimulate the combustion of organic matter. (See Sec. 173.151)
POISON A	6	Extremely Dangerous Poisons – Poisonous gases or liquids of such nature that a very small amount of the gas, or vapor of the liquid, mixed with air is dangerous to life. (Sec. 173.326)
POISON B	6	Less Dangerous Poisons – Substances, liquids or solids (including pastes and semi-solids), other than Class A or Irritating materials, which are known to be so toxic to man as to afford a hazard to health during transportation; or which, in the absence of adequate data on human toxicity, are presumed to be toxic to man. (Sec. 173.343)
IRRITATING MATERIAL	6	A liquid or solid substance which, upon contact with fire or when exposed to air, gives off dangerous or intensely irritating fumes, but not including any poisonous material, Class A. (Sec. 173.381)
ETIOLOGIC AGENT	6	An "etiologic agent" means a viable micro-organism, or its toxin, which causes, or may cause, human disease. (Sec. 173.386)
RADIOACTIVE MATERIAL	7	Any material, or combination of materials, that spontaneously emits ionizing radiation, and having a specific activity greater than 0.002 microcuries per gram. (Sec. 173.389) [See Sec. 173.389(a) through (1) for details]
CORROSIVE MATERIAL	8	Any liquid or solid that causes visible destruction or irreversible alterations in human skin tissue or a liquid that has a severe corrosion rate on steel. [See Sec. 173.240(a) and (b) for details]
ORM – OTHER REGULATED MATERIALS		(1) Any material that may pose an unreasonable risk to health and safety or property when transported in commerce; and (2) Does not meet any of the definitions of the other hazard classes specified in this subpart; or (3) Has been reclassed an ORM (specifically or permissively) according to this subchapter. [Sec. 173.500(a)] NOTE: A material with a flashpoint of 100°F. to 200°F. must be classed as a combustible rather than as an ORM if it is a hazardous waste or is offered in a packaging having a rated capacity of more than 110 gallons.
ORM-A	9	A material which has an anesthetic, irritating, noxious, toxic, or other similar property and which can cause extreme annoyance or discomfort to passengers and crew in the event of leakage during transportation. [Sec. 173.500(b)(1)]

HAZARD CLASS	UN No.	DEFINITIONS
ORM-B	9	A material (including a solid when wet with water) capable of causing significant damage to a transport vehicle from leakage during transportation. Materials meeting one or both of the following criteria are ORM-B materials: (1) A liquid substance that has a corrosion rate exceeding 0.250 inch per year (IPY) on aluminum (nonclad 7075-T6) at a test temperature of 130°F. An acceptable test is described in NACE Standard TM-01-69; and/or (2) Specifically designated by name in Sec. 172.101. [Sec. 173.500(b)(2)]
ORM-C	9	A material which has other inherent characteristics not described as an ORM-A or ORM-B but which makes it unsuitable for shipment, unless properly identified and prepared for transportation. Each ORM-C material is specifically named in Sec. 172.101. [Sec.173.500(b)(3)]
ORM-D	9	A material such as a consumer commodity which, though otherwise subject to the regulations of this subchapter, presents a limited hazard during transportation due to its form, quantity and packaging. They must be materials for which exceptions are provided in Sec. 172.101. A shipping description applicable to each ORM-D material or category of ORM-D materials is found in Sec. 172.101. [Sec. 173.500(b)(4)]
ORM-E	9	A material that is not included in any other hazard class, but is subject to the requirements of this subchapter. Materials in this class include (1) hazardous waste and (2) hazardous substance, as defined in Sec. 171.8. [Sec. 173.500(b)(5)]

THE FOLLOWING ARE OFFERED TO EXPLAIN SOME OF THE ADDITIONAL TERMS USED IN PREPARATION OF HAZARDOUS MATERIALS FOR SHIPMENT. (Sec. 171.8)

CONSUMER COMMODITY (See ORM-D above)		A material that is packaged or distributed in a form intended and suitable for sale through retail sales agencies or instrumentalities for consumption by individuals for purposes of personal care of household use. This term also includes drugs and medicines. (Sec. 171.8)
FLASH POINT		The minimum temperature at which a substance gives off flammable vapors which, in contact with a spark or flame, will ignite. For liquids, see Sec. 173.115; for solids, see Sec. 173.150.
FORBIDDEN		Material is prohibited from being offered or accepted for transportation. This prohibition does not apply if these materials are diluted, stabilized, or incorporated in devices and they are classed in accordance with the definitions of hazardous materials. [Sec. 172.101(d)(1)]
HAZARDOUS SUBSTANCE		For transportation purposes, a material and its mixtures or solutions, that is identified by the letter "E" in Column (1) of the Hazardous Materials Table, Sec. 172.101, when offered for transportation in one package, or in one transport vehicle if not packaged, and when the quantity of the material therein equals or exceeds the reportable quantity (RQ). For details, refer to Sec. 171.8 and Sec. 172.101 (Hazardous Materials Table).
HAZARDOUS WASTE		Any material that is subject to the hazardous waste manifest requirements of the Environmental Protection Agency specified in the CFR, Title 40, Part 262 or would be subject to these requirements in the absence of an interim authorization to a State under Title 40, CFR, Part 123, Subpart F. (Sec. 171.8). Questions regarding EPA hazardous waste regulations, call Toll Free: (800) 424-9065 or in Washington: 554-1404.
LIMITED QUANTITY		The maximum amount of a hazardous material as specified in those sections applicable to the particular hazard class for which there is a specific labeling and packaging exception from the requirements. See Sec. 173.118, 173.118(a), 173.153, 173.244, 173.306, 173.345, 173.364 and 173.391.
REPORTABLE QUANTITY		The quantity of hazardous substance specified in the Hazardous Materials Table (Sec. 172.101) and identified by the letter "E" in Column (1). (Sec. 171.8)

✱ THIS HANDOUT IS DESIGNED AS A TRAINING AID FOR ALL INTERESTED PARTIES WHO MAY BECOME INVOLVED WITH HAZARDOUS MATERIALS. IT DOES NOT RELIEVE PERSONS FROM COMPLYING WITH THE DEPARTMENT OF TRANSPORTATION'S HAZARDOUS MATERIALS REGULATIONS. SPECIFIC CRITERIA FOR HAZARD CLASSES AND RELATED DEFINITIONS ARE FOUND IN THE CODE OF FEDERAL REGULATIONS (CFR), TITLE 49, PARTS 100-177.

Information Services Division, DMT-11
Office of Operations and Enforcement
Materials Transportation Bureau
Department of Transportation
Washington, D.C. 20590

APPENDIX VIII

Preparedness and Prevention

Source: Reprinted from 40 CFR Part 265 Subpart C.

Subpart C—Preparedness and Prevention

§ 265.30 Applicability.

The regulations in this subpart apply to owners and operators of all hazardous waste facilities, except as § 265.1 provides otherwise.

§ 265.31 Maintenance and operation of facility.

Facilities must be maintained and operated to minimize the possibility of a fire, explosion, or any unplanned sudden or non-sudden release of hazardous waste or hazardous waste constituents to air, soil, or surface water which could threaten human health or the environment.

§ 265.32 Required equipment.

All facilities must be equipped with the following, *unless* none of the hazards posed by waste handled at the facility could require a particular kind of equipment specified below:

(a) An internal communications or alarm system capable of providing immediate emergency instruction (voice or signal) to facility personnel;

(b) A device, such as a telephone (immediately available at the scene of operations) or a hand-held two-way radio, capable of summoning emergency assistance from local police departments, fire departments, or State or local emergency response teams;

(c) Portable fire extinguishers, fire control equipment (including special extinguishing equipment, such as that using foam, inert gas, or dry chemicals), spill control equipment, and decontamination equipment; and

(d) Water at adequate volume and pressure to supply water hose streams, or foam producing equipment, or automatic sprinklers, or water spray systems.

§ 265.33 Testing and maintenance of equipment.

All facility communications or alarm systems, fire protection equipment, spill control equipment, and decontamination equipment, where required, must be tested and maintained as necessary to assure its proper operation in time of emergency.

§ 265.34 Access to communications or alarm system.

(a) Whenever hazardous waste is being poured, mixed, spread, or otherwise handled, all personnel involved in the operation must have immediate access to an internal alarm or emergency communication device, either directly or through visual or voice contact with another employee, *unless* such a device is not required under § 265.32.

(b) If there is ever just one employee on the premises while the facility is operating, he must have immediate access to a device, such as a telephone (immediately available at the scene of operation) or a hand-held two-way radio, capable of summoning external emergency assistance, unless such a device is not required under 265.32.

APPENDIX IX

Contingency Procedures

Source: Reprinted from 40 CFR 262.34.

(i) At all times there must be at least one employee either on the premises or on call (*i.e.*, available to respond to an emergency by reaching the facility within a short period of time) with the responsibility for coordinating all emergency response measures specified in paragraph (d)(3)(iv) of this section. This employee is the emergency coordinator.

(ii) The generator must post the following information next to the telephone:

(A) The name and telephone number of the emergency coordinator;

(B) Location of fire extinguishers and spill control material, and, if present, fire alarm; and

(C) The telephone number of the fire department, unless the facility has a direct alarm.

(iii) The generator must ensure that all employees are thoroughly familiar with proper waste handling and emergency procedures, relevant to their responsibilities during normal facility operations and emergencies;

(iv) The emergency coordinator or his designee must respond to any emergencies that arise. The applicable responses are as follows:

(A) In the event of a fire, call the fire department or attempt to extinguish it using a fire extinguisher;

(B) In the event of a spill, contain the flow of hazardous waste to the extent possible, and as soon as is practicable, clean up the hazardous waste and any contaminated materials or soil;

(C) In the event of a fire, explosion, or other release which could threaten human health outside the facility or when the generator has knowledge that a spill has reached surface water, the generator must immediately notify the National Response Center (using their 24-hour toll free number 800/424-8802). The report must include the following information:

(1) The name, address, and U.S. EPA Identification Number of the generator;

(2) Date, time, and type of incident (*e.g.*, spill or fire);

(3) Quantity and type of hazardous waste involved in the incident;

(4) Extent of injuries, if any; and

(5) Estimated quantity and disposition of recovered materials, if any.

187

APPENDIX X

Guide for Reuse of Packagings
(Boxes, Kegs, Cylinders and Steel Drums)

Source: Courtesy of Research and Special Programs Administration of United States Department of Transportation.

The following information has been abstracted from Code of Federal Regulations, Title 49, Parts 100-177 and is intended to serve as an aid for in-house use when reviewing the requirement on the reuse of containers. It does not include or refer to all applicable requirements.

1. REQUIREMENTS (Sec. 173.28)

    A. CONTAINERS – Any container used more than once (refilled and reshipped after having been previously emptied) must meet the Code requirements. That is, containers must be in such condition, that they comply in all respects with the prescribed requirements. This includes container closing devices and cushioning materials.

    B. REPAIR OF CONTAINERS – Repairs to containers must be made in accordance with requirements for materials and construction as prescribed in Parts 178 and 179 of Title 49 for new containers, or as otherwise prescribed. All parts that are weak, broken, or otherwise deteriorated must be replaced.

    C. MARKING AND LABELING
        (1) All markings applied and prescribed by the regulations must be maintained in a legible condition.
        (2) If the prescribed markings cannot be kept plain and legible, then a metal plate, with a reproduction of the prescribed markings plainly stamped thereon may be brazed, soldered or securely fastened to the containers.
        (3) All containers previously used for the shipment of any hazardous materials must have the old markings thoroughly removed or obliterated before being used for the shipment of other articles. These markings include the name of contents, addresses, and labels.

2. USE OF CONTAINERS (Sec. 173.28)

    A. Boxes previously used for High explosives containing a liquid explosive ingredient not contained in an inside metal container must not be used again for shipments of any character.

189

B. Boxes that have been contaminated by liquid explosive composition must not be used for shipment of any character.

C. Kegs previously used for any chlorate must not be used for shipments of any character.

D. Metal Kegs previously used for black powder not contained in any interior package must not be used for shipment of any explosive.

E. Containers used for shipments of etching acid, n.o.s must not be reused for shipment of any commodity.

F. Cylinders used in anhydrous hydrofluoric acid service must comply with the requirements of Sec. 173.264(b)(1) AND must not be used in any other services.

3. REUSE OF DOT SPECIFICATIONS: 17C, 17E AND 17H STEEL DRUMS (Sec. 173.28(m))

A. Specification 17C, 17E, and 17H steel drums which contents have been removed, may be reused as prescribed in Part 173. They can be used as packagings for shipment of flammable liquids, flammable solids, organic peroxides, oxidizers, poisons (see Sec. 173.370, radioactive materials and corrosive liquids (see Sections 173.249 and 173.249(a). However, only use if the following requirements, in addition to other requirements, of Sec. 173.28(m) are complied with PRIOR to each reuse.

NOTE: Containers that do not meet the requirement of DOT specification containers can be reused for Corrosive solids and any other hazardous materials. However, the commodity being packaged must not be capable of reacting with the steel container. The major requirements are outlined below:

1. Visual Inspection - Each drum must be thoroughly cleaned to remove all residue and foreign matter. It must be inspected for deterioration or defects. Parts that are weak, broken or otherwise deteriorated must be replaced. Closure devices and parts must be removed (if removable) and inspected for defects. Each open-head gasket must be replaced. Any drums which show evidence of deterioration such as:

   a. Visible pitting or creases,
   b. Significant reduction in parent metal thickness from rust, corrosion, metal fatigue or other material defect.

   If it cannot be returned to its original shape and contour it DOES NOT QUALIFY for reuse.

   NOTE: All repairs must be made in accordance with requirements for materials and construction as prescribed in the regulations for new containers.

2. Air Pressure Test for Leakage - Except for the removable head and adjacent chime area, the entire surface of each closed-head drum and each open-head drum, must be tested for leakage by constant internal air pressure.

   a. The leakage test must be conducted by (1) submersion under water; (2) completely covering the surface with soap suds or oil, or; (3) some other method that will be considered EQUALLY SENSITIVE.

   b. Leakers shall be rejected or repaired and retested. Repairs must be made by methods used in constructing containers and NOT BY SOLDERING. The air pressure must be maintained for a period of time sufficient to permit a complete inspection for leaks. The minimum constant internal air pressure for testing must be as follows:

Spec. No.	Capacity	Minimun Test Pressure Pounds per square inch (psi)
17C	All	15 psi
17E	Over 12 gallons	7 psi
17E	12 gallons, or less	5 psi
17H	Over 12 gallons	7 psi
17H	12 gallons, or less	5 psi

B. Equally Sensitive Test –

    1. Outlined below are Leakage Test Methods Considered "Equally Sensitive" for Reconditioned 17C, 17E and 17H Drums.

    2. A number of questions have been raised concerning what other test methods would be considered "Equally Sensitive". Any test procedure is considered by the Office of Operations and Enforcement to be as "Equally Sensitve" as the methods specified in Section 173.28(m)(2) of the Hazardous Materials Regulations. If it—

        a. Will subject a drum to constant internal air pressure (at equilibrium with the system closed) at the specified minimum pressure.

        b. Utilizes an accurate pressure gauge or other measuring device which will permit readings to an accuracy of .10 psig (pounds per square inch gage pressure).

        c. Allows for sufficient time to discover leaks; and the process is reproducible.

        NOTE: A visual inspection procedure that does not employ the minimum air pressure specified MAY NOT be used to qualify a drum for reuse under Section 173.28(m)(2).

    3. Other test procedures not meeting the prescribed tests or all of the above "Equally Sensitive" criteria are not considered adequate to meet the requirements of these standards unless specific approval has been obtained from the Materials Transportation Bureau.

        a. Markings – All previous test markings, commodity identification markings, and labels must be removed.

            (1) All drums that qualify for reuse must be marked on the body within 10 inches of the top with the following information:

                a. "Tested"
                b. Month and Year it was Tested.
                c. DOT Registration Number of the reconditioner.

            (2) Markings must be at least 1/4 inch figures and the letters on contrasting background. (See figures 1 and 2)

EXAMPLE: TESTED 2/74
DOT R1000

(DOT 17E)
Tight-head 20/18-gauge 55-gal. drum

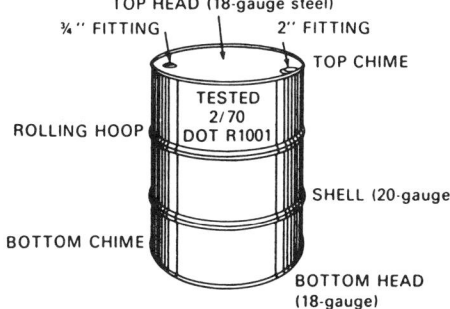

Figure 1 DOT 17E – 55-gal. drum

b.   Any drum meeting one specification which has been altered to meet another specification must be capable of meeting the new specification in all respects. Drums converted to meet another specification must bear the original specification markings. (See Figure 2)

c.   The old and new specification identification in conjunction with the markings shown above are required.

EXAMPLE: 17E/17H
            TESTED 2/74
            DOT R1000

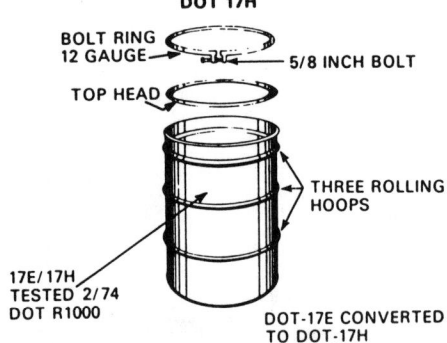

DOT 17H

BOLT RING
12 GAUGE →   — 5/8 INCH BOLT
TOP HEAD

THREE ROLLING HOOPS

17E/17H
TESTED 2/74
DOT R1000

DOT-17E CONVERTED TO DOT-17H

Figure 2 DOT-17E Converted to DOT-17H

d.   The DOT Registration number required for this marking must be obtained from: Associate Director for Office of Hazardous Materials Regulations, Materials Transportation Bureau, Washington, D.C. 20590.

This publication may be reproduced without special permission from this office.

INFORMATION SERVICES DIVISION
OFFICE OF OPERATION AND ENFORCEMENT
MATERIALS TRANSPORTATION BUREAU
DEPARTMENT OF TRANSPORTATION
WASHINGTON, D.C. 20590

REVISED JULY 1983

APPENDIX XI

Federal Regulation (40 CFR 261.7) Regarding Hazardous Waste Residues in Empty Containers

§261.7 RESIDUES OF HAZARDOUS WASTE IN EMPTY CONTAINERS.

(a)(1) Any hazardous waste remaining in either (i) an empty container or (ii) an inner liner removed from an empty container, as defined in paragraph (b) of this section, is not subject to regulation under Parts 261 through 265, or Part 270 or 124 of this chapter or to the notification requirements of section 3010 of RCRA.

(2)Any hazardous waste in either (i) a container that is not empty or (ii) an inner liner removed from a container that is not empty, as defined in paragraph (b) of this section, is sub-

ject to regulation under Parts 261 through 265, and Parts 270 and 124 of this chapter and to the notification requirements of section 3010 of RCRA.

(b)(1) A container or an inner liner removed from a container that has held any hazardous waste, except a waste that is a compressed gas or that is identified as an acute hazardous waste listed in §§ 261.31, 261.32, or 261.33(e) of this chapter is empty if:

(i) All wastes have been removed that can be removed using the practices commonly employed to remove materials from that type of container, *e.g.*, pouring, pumping, and aspirating, *and*

(ii) No more than 2.5 centimeters (one inch) of residue remain on the bottom of the container or inner liner, *or*

(iii)(A) No more than 3 percent by weight of the total capacity of the container remains in the container or inner liner if the container is less than or equal to 110 gallons in size, or

(B) No more than 0.3 percent by weight of the total capacity of the container remains in the container or inner liner if the container is greater than 110 gallons in size.

(2) A container that has held a hazardous waste that is a compressed gas is empty when the pressure in the container approaches atmospheric.

(3) A container or an inner liner removed from a container that has held an acute hazardous waste listed in §§ 261.31, 261.32, or 261.33(e) is empty if:

(i) The container or inner liner has been triple rinsed using a solvent capable of removing the commercial chemical product or manufacturing chemical intermediate:

(ii) The container or inner liner has been cleansed by another method that has been shown in the scientific literature, or by tests conducted by the generator, to achieve equivalent removal; or

(iii) In the case of a container, the inner liner that prevented contact of the commercial chemical product or manufacturing chemical intermediate with the container, has been removed.

[45 FR 78529, Nov. 25, 1980, as amended at 47 FR 36097, Aug. 18, 1982; 48 FR 14294, Apr. 1, 1983; 50 FR 1999, Jan. 14. 1985]

Index